Think Like a Lawyer about...

Artificial Intelligence

Theodore F. Claypoole

ABA
AMERICAN BAR ASSOCIATION
Business Law Section

Cover design by Catherine Zaccarine.

Printed in the United States of America.

29 28 27 26 25   5 4 3 2 1

ISBN: 978-1-63905-608-8

Table of Contents

Foreword

This book is for lawyers, policymakers, executives, and all other people who want to understand artificial intelligence (AI). There are dozens of accurate ways to describe and define AI, but few capture its full impact on society. I titled the book "Think Like a Lawyer about Artificial Intelligence" because lawyers are forced to focus on how a technology interacts with our civilization and on the risks that technology causes for their clients, themselves, and the systems they operate within. We can all benefit from thinking about how this crucial technology functions in the real world and how it alters societal risks.

I understand that many of the specific examples in this book will quickly become old news and the cases cited for legal discussions will be decided, overturned, or superseded by others post-publication. This should not negatively affect future readers of this book. In nearly all chapters, cases and laws serve as examples of certain types of thinking, rather than support for a "correct" analysis. We are exploring together how lawyers think about risk, about technology, and about how rules affect us all. When those winds shift, it only proves my point that the wind exists and that the wind, whichever direction it blows, affects our behavior. "Thinking like a lawyer" is an eternal exercise. This book explores legal thinking methods through a crashing wave of new technology swamping our previously comfortable interactions. Once the wave has subsided, leaving its imprint on all that it touched, the analysis is still valid, and it can help us anticipate and manage the next wave.

To that end, I advise thinking about AI through its practical functionality, analyzing how AI is actually used in day-to-day life and how those operations create risk. I have assigned these functional groups to incorporate differing sets of legal issues, so that the groupings are understandable as distinct legal and policy problems. Many of these functional groupings use nearly identical underlying technology and their development may be indistinguishable from each other to technologists. In addition, it is likely that if this book is updated or rewritten five years hence, entirely new functionalities will have arisen and a different set of categories would be more appropriate, especially as generative AI splits into multiple silos of usage. However, none of these changes affects the basic premise of this book—that people would benefit from understanding AI as a series of functionalities impacting our society in different ways. This helps us understand not just the technology itself, but its significance for all of us.

Lawyers are called to help with real-world problems. Sometimes these problems involve technology or are created by the new applications that a technology can produce. One of the most vexing problems addressed in this book involves deepfakes of real people generated by AI. Whether pornographic deepfakes or those generated to shed bad light on political candidates, people's lives can be ruined through this process, and the only practical solution is application of the law. In some cases, new laws are needed. In others, current rules protect the innocent. But attorneys are asked to think creatively about solutions for novel disasters that new technology spawns.

Policymakers must also consider impact and risk. Initial panic about the negative possibilities of generative AI led many public voices to advocate for sweeping regulation of AI technology. Some cited a failure to limit the societal damage of social media a decade earlier as evidence that AI should be hog-tied as a general matter to prevent harms that we could predict but which had not become widespread. Such thinking is destructive when AI, in its many forms and functions, can also remake our society in positive ways. This book advocates for precise thinking, precise definitions, and precise regulation of known harms, rather than wholesale restriction of useful technology.

So as the law shifts, various types of AI advance in utility and functionality, and the world learns more ways to incorporate the AI into lives, we should still try to think like a lawyer about AI.

Acknowledgments

Thank you to William Carter, Tyler Conner, and Smita Rajmohan for your contributions to this book.

CHAPTER 1

What Is Artificial Intelligence?

In 1956, when even the smallest computers filled entire rooms with tubes and wires, Massachusetts Institute of Technology (MIT) professor John McCarthy described the attempt to replicate human thinking on machines as "artificial intelligence." Only six years earlier Alan Turing and John von Neumann helped create the modern computing age by switching the underlying basis of computations from a decimal system that had been used since the 1800s to the binary system of 1s and 0s that we use today. Turing asked, "Can machines think?" Even early in the process of inventing computing power, academics were considering how these new machines could perform human tasks.

The term "artificial intelligence" or "AI" has traditionally been aspirational, not describing computing capabilities we have, but describing what we expect those computers to evolve into. Science fiction writer Ted Chiang has defined artificial intelligence as "a poor choice of words in 1954," preferring instead to call our current technologies "applied statistics." Former Columbia University Computer Science professor Eric Siegel was more blunt. He said, "AI is a big fat lie. Artificial intelligence is a fraudulent hoax—or in the best cases it's a hyped-up buzzword that confuses and deceives." Siegel is an advocate of the technology, but prefers the label "machine learning."[1]

When we use the term artificial intelligence to describe existing technology, we are surrendering to our natural impulse to humanize

1. Siegel said this on his YouTube show *Dr. Data*. Transcript of Eric Siegel, *Dr. Data Video: A.I. Is a Big Fat Lie*, Machine Learning Times, Jan. 22, 2019 (Login required).

everything. We project human-like thoughts and emotions on our pets, our smartphones, our vehicles, even rivers and oceans. Ancient (and some modern) people identified supernatural beings as the controllers of natural phenomena; it was easier to understand volcanic or tectonic activity by ascribing personal anger or retribution as the cause. But when we apply human motivations or thought processes to non-human actors, we can deeply misunderstand what is really happening behind the scenes—we misidentify the context and intention. A truck doesn't want to move forward when you put it in first gear, it is compelled to do so. A parrot doesn't know the meaning of all the words it can repeat.[2] A generative AI model can draw a picture in the style of Rembrandt, but it doesn't intend to create a great work of art. It has no aims at all.

A computer is not thinking or intending. It performs tasks determined by inputs and programming. We would be wrong to think that humans act in exactly the same way. None of us is rational, and we don't even make calculations and decisions with merely brainpower. Melanie Mitchell, the Davis Professor of Complexity at the Santa Fe Institute states, "What we've learned from research in embodied cognition is that human intelligence seems to be a strongly integrated system with closely interconnected attributes, including emotions, desires, a strong sense of selfhood and autonomy, and a common sense understanding of the world." She adds, "It's not at all clear that these attributes can be separated."[3] But these are attributes not shared with machines of any type, which do not have emotions, desires, common sense, or selfhood. I state below that computers should not receive human rights until the computers can dream. Our system of laws is built around rewards and punishments that only make sense when applied to a being with desires comparable to humans and a similar limited lifespan. Why would we grant those rights to a computer if it had no use for them? When a computer performs a task that we can also perform, we tend to assume the machine is making the same judgments and decisions in its binary processors that we are making when we perform the task. It isn't. It can't.

---

2. Although the parrot may be noting the reactions of those around her when she makes a certain sound.

3. The Physics arXiv Blog, *The 4 Fallacies of Artificial Intelligence: Artificial intelligence researchers are kidding themselves that human-level performance is within reach, argues one leading thinker. Here's why*, DISCOVER MAGAZINE, May 1, 2021, quoting Melanie Mitchell, *Why AI is Harder Than We Think*.

If you ask generative visual-based AI to create a photo-realistic picture of a father and daughter throwing a softball between them, many of today's models can produce this for you. But if you ask the same AI if the softball can be thrown to the moon, if it dissolves in raindrops, if it can be stretched to one hundred times its size, or can be eaten by people, the AI will not know without additional research. It has no context. It can only predict what a softball should look like by comparing and contrasting to labelled pictures. Performing a complex task is not thinking in the way we understand it. The cotton gin, patented in the United States in 1794, could perform an economically valuable task much better than humans—it could quickly separate strands of cotton from their seeds. Creating the cotton gin involved brilliant practical conceptualization by Eli Whitney, but operating the machine, the process of separating cotton fibers from seeds, is mechanical and involves no thinking. We don't ascribe brilliance to the machine itself because we can see how it works. If more people understood the way that sophisticated algorithms or machine learning models work, they would be less likely to believe the models to be intelligent.

One of the most significant divisions in the artificial intelligence world is the difference between "narrow AI" and "general AI." Narrow AI is a computational system that is developed and directed to solving one specific problem or set of tasks. In many cases, the narrow AI can outperform humans in the specific task that it specializes in. Some narrow AIs detect cancer by reading pictures created by x-rays. Some narrow AIs predict hazardous weather events. Some narrow AIs can beat you at chess. But the narrow AI cannot beat you at chess, secure tickets for a concert, and drive your car. This would be the role of a general AI, and general AI doesn't exist.

We currently have narrow AIs that seem to be performing many disparate tasks, but those tasks tend to be related to the specific training of the AI model, so the scope of its expertise is still narrow. We have huge AI models like IBM's Watson that beat the champions at *Jeopardy!* and was touted as a solution-finder for business and social problems. It turns out the AI platform couldn't perform as promised, and the company has scaled back its AI aspirations.[4] We now have generative AI, which seems

4. "David Ferrucci, the scientist, explained that Watson was engineered to identify word patterns and predict correct answers for the trivia game. It was not an all-purpose answer box ready

to develop answers to all kinds of problems, but which generates words or pictures without any context or understanding. If we read the story or view the picture and automatically believe that its creator had the same or better grasp of the context than a human would hold, then we would be wrong about the AI.

"Narrow AI" is not intelligence in the way that people understand it, simply an ability to perform the set of tasks the machine was trained to perform. If a parrot can recite Lady Macbeth's soliloquy, the parrot has not become intelligent in the manner of a human adult. The parrot can perform the task it was trained to do.[5] AI models can run calculations at mind-blowing speed and often have access to enormous amounts of computing power minute-to-minute, so they can combine tasks that are truly complicated, and perform many tasks much faster and more accurately than humans can. On the other hand, bulldozers are also tools successfully designed to vastly outperform humans in their chosen tasks, but we do not despair of the possibility that bulldozers will control our world.[6] Neither computers nor bulldozers have the essential context or thinking power that we expect from a four-year-old human and neither tool can perform at all outside of its trained task range. Humans can think about solutions to problems of many different types across all the situations we encounter. This is general intelligence. It allows us to apply thoughts and contextual considerations from one arena to any other. So far, the only AI general intelligence we have met lives in fiction. No current AI has anything close to general intelligence.

Bulldozers are built for a human operator, while some AI models can improve themselves without further human intervention. Shouldn't we be more concerned about the computers? AI models seem to learn faster and easier with supervised learning, where a human uses annotated data to show the AI model which picture is a cat and which is a dog. However, AI models can learn in an unsupervised fashion as well. Some of this training

to take on the commercial world, he said. It might well fail a second-grade reading comprehension test." [Steve Lohr, *What Ever Happened to IBM's Watson?*, N.Y. TIMES, Jul. 16, 2021.]

5. This is not a judgment on whether parrots or other non-human animals have intelligence, which is the topic for other books. Instead, this is simply an observation that being able to regurgitate text from memory without understanding the words is not an act of intelligence.

6. AI-managed bulldozers are a possible option for the next Zack Snyder movie, but as unrealistic as the haunted car in *Christine* or the cyborgs in *The Terminator*.

is conducted through reinforcement learning, where the AI model is presented a task and it applies trial and error until it reaches a way to complete the task. Thus far, self-supervised learning has eluded AI models, which are still simply directed by humans in narrow task-ranges.

AI models may be able to run complex systems and produce impressive artworks and text documents. How is this not intelligent? Generative AI models are prediction machines. They are predicting which word, code, or pixel should come next to meet the prompt. Predicting what a Rembrandt portrait of a submarine captain would look like or how Hemmingway might write a story about space travel can produce some amazing results, but such predictions are still not "thinking" in the way that we would define it.[7] AI, as a super-predictor, can be accurate, while calling that activity super-intelligent is not accurate. It may be tempting to ascribe human characteristics to performing a task we think of as human, but the tasks now tacked by AI can be performed without intelligence.

We have a tendency to overestimate AI, but conversely, we shouldn't fool ourselves into thinking that AI is harmless. AI models can be impressive, dangerous, and important to our society. As stated in the *Harvard Business Review*, "AI Won't Replace Humans—But Humans With AI Will Replace Humans Without AI."[8] The types of algorithms, models, and technologies that we are currently calling AI are already changing business, government, the military, and society. AI, when trained on databases of personal information, can help a bank decide whether you are likely to either be a helpful employee or an unacceptable credit risk. AI, when attached to a database of human faces, can pick a person out of a peaceful crowd, name the person, and suggest where she could be found. AI, when given control of a weaponized drone, could kill a terror suspect. Generative AI is currently making written, photographic, and auditory evidence questionable as it creates realistic-but-fake documents at people's request.

---

7. As the technology has become more embedded in everyday applications, however, there's been growing interest in neuromorphic engineering, an area of research that seeks to emulate the human brain's architecture through the design of specialized hardware and algorithms optimized for low power consumption and real-time processing." This would make AI more brain-like, but would not make it think.

8. Interview with Harvard Business School professor Karim R. Lakhani, co-author of *Competing in the Age of AI: Strategy and Leadership When Algorithms and Networks Run the World*, conducted by Harvard Business Review editor in chief Adi Ignatius, published on August 4, 2023.

All of these current uses of AI could harm us personally and change society.

This book argues that we need to understand how AI is being used so that we can address the societal problems those activities will cause. As stated in a Brooking Institution commentary, "Because AI is a multifaceted capability, 'one-size-fits all' regulation will over-regulate in some instances and under-regulate in others. The use of AI in a video game, for instance, has a different effect—and should be treated differently—from AI that could threaten the security of critical infrastructure or endanger human beings. AI regulation, thus, must be risk-based and targeted."[9] It is not enough to simply say "AI is a menace and must be fitted with guardrails" given the enormous breadth, scope, and reach of AI in our society. A more precise understanding is also important because AI is not implementing itself—people are finding the functionality of AI models to be helpful in their work and personal lives. As AI continues to perform valuable functions, we should not encourage a broad banning of the tool simply because the tool may be used for bad reasons.

So, the self-knowing, all powerful AI of fiction is not close to a reality. Just because a machine (or animal) imitates human language or behavior—or finds a novel method of achieving assigned goals—does not mean that the non-human is thinking like you and I do. We have not created AI that will run its own business or small town, despite creating software simulations for running mock digital businesses or small towns. However, our current AI could definitely be dangerous, either here in the physical world operating military drones or driverless taxis, or by making decisions about our life such as whether we qualify for government-subsidized housing. Current AI can identify or misidentify us as present at a crime scene. It can place our face in a picture or video that would be deeply misleading, but still humiliating, and believed by most people. Even AI created for narrow functionalities can harm us.

So, what is current AI, if not a machine version of human thought? It is, instead, a software model that has been training to perform a task or set of tasks, and then continually refined with more input and direction. Some machine learning models have been given little direction, and some have worked primarily with labeled data. A generational AI program may

9. Tom Wheeler, *The three challenges of AI regulation*, Brookings Commentary, Jun. 15, 2023.

have been taught all the rules of the English language, and it may be able to predict the most likely answer to a specific set of questions—like IBM's Watson was taught to win *Jeopardy!*—but that doesn't mean the program understands why a question should be answered a certain way or whether Toronto is a city with both a lake and an airport. The AI seems like it understands English, and for many purposes, it can provide "correct" responses. It is predicting and sorting potential responses, not thinking or understanding.

According to Professor Melanie Mitchell, machine learning models cannot generalize or exhibit common sense "because such systems are susceptible to shortcut learning: learning statistical associations in the training data that allow the machine to produce correct answers but sometimes for the wrong reasons. In other words, these machines don't learn the concepts we are trying to teach them, but rather they learn shortcuts to correct answers on the training set—and such shortcuts will not lead to good generalizations. They are like students who memorize answers to a test without learning why those answers are correct.

Indeed, deep learning systems often cannot learn the abstract concepts that would enable them to transfer what they have learned to new situations or tasks."[10] Mitchell observes that not only our humanizing language encourages false expectations of complex AI, but there are other clear fallacies as well, such as the fact that people naturally see narrow intelligence on a continuum with general intelligence or that the things that are easy for people would be easy for machines.

Doctors have been experimenting with artificial hearts since the 1940s, and a medical team successfully implanted the Jarvik-7 artificial heart in a human in 1982. We have an intimate understanding of how the mammalian heart operates in part, because the heart is primarily a mechanical organ operating on electrical stimulation. It is easier for us to understand and manipulate than other organs like the liver or pancreas. It is important for this discussion to note that humans are still a very long way from understanding the way that brains work, especially the alchemy that leads human brain cells to support cognitive functions. As we get deeper in our knowledge of brain pathologies and baseline electrical operations, researchers peel away layers of mystery about brains, but

10. Melanie Mitchell *supra* note 3.

we still have no agreed general theory of how these cells support a mind, a personality, a set of emotions, or a sense of self. As we learn more, we also find that other parts of a human's physical being, from its breathing to the intestinal biome, can affect the operations of the brain. As observed recently by Professor Hannah Thomasy in the *MIT Technology Review*, "There is no greater scientific mystery than the brain. It's made mostly of water; much of the rest is largely fat. Yet this roughly three-pound blob of material produces our thoughts, memories, and emotions. It governs how we interact with the world, and it runs our body."[11] Neil deGrasse Tyson has said, "Everything we do, every thought we've ever had, is produced by the human brain. But exactly how it operates remains one of the biggest unsolved mysteries, and it seems the more we probe its secrets, the more surprises we find."

How does the brain allocate resources? How are memories stored and retrieved, and why is there an apparent difference between the retrieval of short-term memories and long-term memories? Why does the brain use 20 percent of the body's oxygen and what happens in the brain's unstimulated baseline activity? What does sleep do for us, precisely, and how does this work? Why and how does our brain allow us to hypothesize and run simulations of situations which haven't occurred yet? How do portions of the brain specialize and then integrate those specialized functions? What are emotions and how do they relate to conscious thought? Most importantly, what is intelligence and what is consciousness? It seems that a basic understanding of all of these questions would be important to undergird an attempt to model our own thinking or to make a machine that thinks like we do.

How do we expect to create an "artificial intelligence" when we don't even know how our own intelligence works? We understand logic—an artificial structure imposed on thinking—and can ask computers to apply this logic. And we are skilled at building limited games within defined parameters, and it turns out that computers, with their parallel processing speed and ability to hold nearly unlimited data in a version of memory, can perform better at these limited games than we can. If we want a true artificial intelligence, we should probably understand more deeply how

11. Hannah Thomasy, *Eight ways scientists are unwrapping the mysteries of the human brain*, MIT Tech. Rev., Aug. 24, 2021.

our own thinking works and how our thinking is influenced and limited by biology. We may never be able to create a digital version of our general intelligence or common sense, but we are unlikely to replicate our thinking until we understand the basic processes that cause and influence that thinking. Our current science is far from this goal.

So if the computer activity that is informally known as "artificial intelligence" isn't a mechanized version of human intelligence, what is it? The answer depends on who you ask. AI designers and software programmers will often have a technical answer based on how an AI model is created. They would say that, unlike standard software algorithms, an AI model is developed by feeding huge amounts of information into a specially-designed algorithm, resulting in a computer model that can react to stimuli or prompts and produce results influenced by the information the model was fed. IBM, on its website, takes this position, stating, "At its simplest form, artificial intelligence is a field, which combines computer science and robust datasets, to enable problem-solving. It also encompasses sub-fields of machine learning and deep learning, which are frequently mentioned in conjunction with artificial intelligence. These disciplines are comprised of AI algorithms which seek to create expert systems which make predictions or classifications based on input data." So by feeding hundreds of millions of chess games into the specialized algorithm, you will get a model that "has been taught chess" and can play the game against a human or against other AI chess models. By feeding a billion works of art into a specialized algorithm, and including labels of the artists who created each work and what the art represents,[12] you will get an AI model that will produce a picture of an anthropomorphized bison enjoying a cocktail party generated in the artistic style of Jan van Eyck or the German expressionists.

Some define artificial intelligence as machines that can learn. These AI models go beyond their training and continue consuming more data and refining responses to stimuli—new chess games, new road conditions, or new prompts—to learn to be better at the tasks the model was initially trained to perform. Much of the technology referred to as AI today, from large language models running the most disruptive generative

12. Including labels as metadata for the information fed to such systems—"this is a picture of a cat"—is called supervised learning.

AI programs to medical learning models that can identify signs of disease in patients, are built from machine learning technology.

Others define artificial intelligence by how close the AI model's output seems to human tasks and how complex the tasks would be to complete. So calculating out dozens of digits of pi may not seem so intelligent, although computers can do this faster, longer, and deeper than any human. However, driving a vehicle on human roads in the real world seems both complicated and very human-like, so the interworking algorithms that perform this task are thought of as AI.

People's fundamental misunderstanding of AI can lead to bad policy choices. We can overestimate the human characteristics of AI models and fear them as existential threats to society. We can also underestimate AI's ability to perform many tasks much better than people can, which could lead to criminals or terrorists capitalizing on AI competence. This book catalogs the many ways that AI-driven models interact with society, and the policy/legal choices that we can make to maximize their utility and minimize their risks. At the moment, AI is like a Swiss Army knife, containing tools of many features and functions. We should not treat each tool as if it were the same as the others.

Plenty of tools are core constructors of our society—from hammers and backhoes to word processors and spreadsheets—but can also be used for terrible actions. Hammers and backhoes can inflict trauma on people and can destroy valuable property. Word processors and spreadsheets can facilitate fraud. When developing rules to mitigate or discourage these harms, we don't outlaw or even "place guardrails" around the tools. No, we act to specifically limit the kind of threats that worry us, regardless of the tool that is used. We have laws against striking people with blunt objects, not laws against hammers. We have laws against committing financial fraud, not rules against spreadsheets. Similarly, if we are concerned about how AI affects our society, we should be regulating the harms that concern us, not the tools that might cause that harm in the future.[13]

---

13. Note that certain types of activities enabled by complex modern tools are highly regulated, both by demanding minimum safety standards for tools like airplanes or cars that can create great societal and physical risks, and by insisting that humans be ultimately responsible for the operations of these risky vehicles. This book is not intended to provide specific recommendations for making policy around AI, but policymakers may want to consider these successful regulatory structures developed for complicated human tools with risks arising from extension of physical

In addition, a broad technology-restricting approach to regulation is particularly ill-suited to the speed of AI development. AI models are already deeply embedded in nearly every aspect of society from your Amazon Prime orders to the medical evaluation of your x-rays. Tens of thousands of AI developers are creating, growing, improving, and implementing new AI-based solutions to business and social problems as you read this. Determining "what is AI" at any snapshot moment will not reflect the pace of change and the difference between today's AI solutions and those offered three months from now. On the other hand, policymakers could stay ahead of the technical growth by identifying the problems that AI functions and impacts could cause, and developing policy to fix those problems.

Finally, we must remember that human society is large and complicated, and that technology, once discovered, becomes the legacy of us all. If we are not precise in the manner of regulating the technologies that we call artificial intelligence, we may cede the future benefits of those technologies to other societies who are not as quick to implement broad restrictions. China is already investing tens of billions of yuan into developing AI models for all aspects of society, from military to business, to law enforcement, to civilian life. An unwise set of restrictions limiting the development of AI models in the West could provide an advantage to other cultures who may one day use that AI advantage against us. Sam Altman, the current CEO of OpenAI commented about the eagerness to restrict AI models expressed by European regulators, stating "We will try to comply, but if we can't comply, we will cease operating [in Europe]." Unlike social media companies, whose revenues depend completely on the number of users, companies that develop AI are less sensitive to assuring a warm welcome in all jurisdictions. They may be able to provide the latest developments to countries who accept those developments and may not to countries who make broad restrictions of the technology.

---

size, speed, and elevation—allowing humans and their machines to move more quickly, in more places, and with more weight in motion than would otherwise be possible. Our rules essentially make sure these machines work correctly and that they are under the focused control of humans when being used.

If AI subjects us all to physical and societal risks, similar controls may be advisable.

This book does not recommend turning a blind eye to the potential harms of this new and disruptive technology, but only taking a legal and policy stance focused on limiting harms, rather than a position of broader limits on the underlying technology. I believe we can achieve our goals of reducing social harms with more precise thinking about how to protect our society from those harms.

CHAPTER 2

Why Would Lawyers Need to Think Differently about AI?

The previous chapter demonstrated that AI can be defined and considered in a variety of ways. Many of us simply marvel at the amazing breadth of capabilities that the technology exhibits. Some of us fear the power and change it promises. Most of us want to know how it can help us.

In many cases, people's view of AI depends on their relationship to AI. Like the famous fable of the blind men and an elephant, where each man touched a different part of the elephant and believed he was touching a rope (tail), a curtain (ear), or a wall (body), people tend to understand AI through the lens of their jobs. For illustrative purposes, we examine how technologists and business planners define AI around their perspectives. Then, we examine the lawyers' unique perspective and why the lawyers' view of AI is valuable.

Technologists tend to see every advancement in computing as an opportunity to make better tools and solve more complicated problems, and AI or machine learning is no different. The fact that generative AI can generate operational software code on demand—even for difficult technical problems—can both provide a helpful tool to programmers but also give them pause about the long-term safety of their jobs. As a writer (and English major) recently described in *The Atlantic*, in the age of AI, computer science is no longer a safe major. She writes, "In the ultimate irony, software engineers helped create AI, and now they are the American workers who think it will have the biggest impact on their livelihoods, according to a new survey from Pew Research Center. So much for learning to code."[1] However, coders I have spoken to remain optimistic about how AI will change their jobs. Often, discussions of generative AI's effect on coders are consistent with this statement in *Forbes*, "While programmers definitely should not fear AI, they should stay abreast of trends and collaborate closely with it. AI will not replace people; rather, those who can efficiently use AI will replace those who do not want to or cannot do so."[2] So like many of us in the problem-solving fields, people who develop computer code are beginning to understand that embracing AI is likely to be important for their careers.

Technologists can also define AI by its methods of creation. The Technopedia states "Today's AI predominantly operates on conventional CMOS-based hardware[3] and incorporates a blend of traditional algorithms and data-driven machine learning models."[4] When you make bread for a living, you think about the ingredients that go into it and the processes for creating it. Similarly, when you develop AI models for a living, you likely think about that technology as a product and its process—a summary of how it was made and what it was made with. As I write this, the software company SAS explains AI on its website as, "Using deep learning and natural language processing, computers can be trained to accomplish specific tasks by processing large amounts of data and recognizing patterns in the data." So, for some software experts, AI is defined not by what it does, but by the way it was created. An AI can be created using machine learning, deep learning, directed learning, or natural language processing, while a non-AI algorithm can be created with standard human code-writing processes.

A similar technologist's manner of defining AI would be identifying types of logic that the algorithm uses to solve problems and assigning the AI designation to machines that process information differently than standard software. For example, the chess AI models that finally beat human chess masters processed millions of chess games and apparently applied strategies recognized from winning sides. This was different from previous chess programs which built pre-determined strategies into the

1. Kelli María Korducki, *So Much for 'Learn to Code'*, The Atlantic, Sept. 26, 2023.

2. Robert Strzelecki, *Does Artificial Intelligence Threaten Programmers?*, Forbes, Aug. 23, 2023.

3. Referring to the Complementary Metal-Oxide Semiconductor used in computer processors and memory chips.

4. https://www.techopedia.com/definition/190/artificial-intelligence-ai, accessed Sept. 2023.

algorithm. Older programs, some built on a brute-force tree search which tested as many variations of moves possible on a decision tree, could win chess games but for years they could not beat human chess masters. This changed in the late 1990s, but brute force had become so strong—so many variations tested—that a chess bot could ostensibly beat anyone. According to a chess computer history article from *Ars Technica*, "the 'Deep Blue' system that IBM commissioned to challenge [chess grand master] Gary Kasparov could consider one hundred million chess positions per second. That was approximately nine to eleven moves forward. In 1996, Deep Blue beat Kasparov in several games, but the Grandmaster eventually won the match in a score of four to two. In 1997, Kasparov lost a rematch against the machine."[5] But today's developers wouldn't consider Deep Blue to be artificial intelligence, because it just applied standard decision-tree logic at an incredibly fast pace.

It took nearly another two decades for a computer system to beat a 9-dan [top level] master of the game Go. Go is a much harder computing problem than chess because of Go's much larger set of move choices and the game's strategic nature. The computer model that finally beat a 9-dan Go master in 2016 is named AlphaGo, and it used knowledge acquired by artificial neural network machine learning combined with a Monte Carlo tree search algorithm. Professional software and AI coders are likely to define AlphaGo as AI due to its deep learning training and the chance-based elements in its logic. So the methods of training and types of logic are important elements for many experts to differentiate between AI and standard software algorithms.

Many non-tech companies that use AI have tended to see the technology as an optimization machine, as tools for optimizing processes, procedures, and technologies. From the early days of enterprise resource planning systems in the 1990s, business has spent resources on technology to build efficiencies. Some business systems are so complicated, that only a powerful computer model can propose effective strategies to make it work better. Think about the package delivery system at a company like Amazon or United Parcel Service (UPS), with dozens of warehouses, hundreds of airplanes, and thousands of trucks finding the most efficient manner

5. Mathew Lasar, *Brute force or intelligence? The slow rise of computer chess*, ARS TECHNICA, Aug. 5, 2011.

to bring an item you have ordered to your home. People can design an efficient system, but it takes billions of data points running in millions of simulations to understand what basic structural changes can optimize deliveries over the long run. If you are investing in building the infrastructure for such a system, AI tools can provide the type of data crunching and delivery models you need. We live in a complex world, and AI has helped business and government manage it better. So if you ask a UPS logistics director about AI, they will likely to talk about optimization through machine learning models. Complicated processes abound in business and government, from precision manufacturing to military logistics and from attracting consumers to your brand to meeting governance requirements, AI has a niche in making enterprises run better and cheaper.

Enterprises now spend significantly on AI tools both to control data more efficiently and to draw inferences from that data. The industries of data informatics—applying data-centered approaches to solve complex problems—and data analytics—analysis that uses quantitative and qualitative techniques to look for trends and patterns in the data—are growing quickly in the business and government world and are increasingly driven by AI. The Harvard Business School course website lists several examples of how businesses use AI for data analytics, including the following:

> Ensuring a quality user experience is a top priority for ride-hailing company Uber. To streamline its customer service capabilities, the company developed a Customer Obsession Ticket Assistant (COTA) in early 2018—a tool that uses machine learning and natural language processing to help agents improve their speed and accuracy when responding to support tickets.
>
> COTA's implementation delivered positive results. The tool reduced ticket resolution time by 10 percent, and its success prompted the Uber Engineering team to explore how it could be improved.[6]

While businesses and government entities have been caught up in the generative AI frenzy of 2023, AI had previously cut deep inroads into their budgets with other functional aspects of AI. If you ask a process manager

6. Matt Gavin, *Business Insights*, HARV. BUS. SCH. ONLINE, Jan. 15, 2019, https://online.hbs.edu/blog/post/business-analytics-examples.

or data specialist at a large corporation in the last five years what AI means to them, they would likely discuss process optimization and data analytics rather than generation of documents. Because they have been working with AI in important company-changing ways, they would define AI differently than many others.

Lawyers tend to think differently from other people. The essence of a lawyer's job is risk management and advising clients on the perils involved in situations, actions, and decisions. Often that risk counseling is based on understanding the rewards and penalties in the circumstances facing the client. What could be lost—money (how much?), freedom (for how long?), child custody (on what terms?)? What could be gained—money (how much?), a business brand (in what jurisdictions?), access to water (how much and for how long?)? Smart clients consult their lawyers before trouble arises because managing risk is less expensive and more effective when the problems are not yet urgent. The world presents risk and reward in every personal, business, or policy decision, and lawyers are trained to evaluate them. Medical doctors also advise on risk, but they do so in a very specific set of contexts—how will your physical or mental health be affected? Accountants advise about financial risks. Lawyers tend to tackle societal risks based on how people operate their lives within the structures that our world has provided. Attorneys analyze the biggest picture frame available, and apply it to narrow confines of a client's life or business.[7]

So this is one of ways that lawyers approach artificial intelligence. A lawyer asks, "What are the risks?" As with any technology, the AI risks that matter to an attorney are those that can affect a client's life or business. To understand the risks, a lawyer needs to know enough about how a technology works and how it will be used by the client. That does not mean that all lawyers need to be able to engineer the technology themselves in order to advise on it. For certain patent lawyers, being able to build and operate an AI model may be crucial for protecting the most

7. Other professionals in our society also specialize in advising people about risk. For example, priests, pastors, rabbis, and other spiritual advisors help people understand the risks of certain choices on the person's spirit, psyche, or soul, and they join therapists in advising constituents/patients on how actions or inactions can affect mental health and personal lives. Nearly every aspect of life comes with a set of professionals who can provide advice, help reduce risks, and produce better results, from sports psychologists to travel agents, and from nutritionists to interior designers.

valuable inventions with the technology. But for business and personal lawyers, it is most important to know how the AI will be used and what risks does this functionality bring to the client. This is one of the reasons that lawyers tend toward a functional categorization and approach to AI models. Understanding function is important to understanding risk.

Another mark of legal thinking is a knowledge and application of laws, rules, regulations, and restrictions of all type. Some of those restrictions may be imposed by society and some may be voluntarily accepted in contracts. Lawyers are trained in specialized schools and through years of practice to understand the applicability of limitations and the costs for breaking the rules. The laws are a barrier only if a person agrees to obey them. Sometimes a client is willing to pay the cost of ignoring the rules altogether. The mobile ride application market dominated by Uber and Lyft would not exist today if those companies had decided to completely comply with the taxi laws that stood during their early years. The ride-sharing companies made a legal decision that fighting these laws was worth the financial risks, and this risk changed the way that professional driving services work around the world.

Most famously, lawyers are experts in conflict. Many spend their work lives in the lower-intensity conflict of deal-making and contract negotiations. These lawyers help clients push for the best and most favorable business relationships as services are purchased and entire companies are acquired. They must aim to reach a conclusion where all parties are comfortable moving forward. Some lawyers build a career from fighting government agencies to gain benefits for their clients, or fighting private parties on behalf of a government agency. Specialists at understanding the administrative workings of a regulator or benefits manager who often acts as both judge and opposing party, these lawyers fix conflicts over taxes, government payments, and intellectual property protections. A specialized group of these lawyers are lobbyists and legislators, who bring policy fights to the halls of Congress, Parliament, and state assemblies. Policies and budget advantages are the usual targets driving the conflicting parties in legislative discussions.

But it's likely that many of us think of battling litigators as the most iconic sources of legal conflict. Some of my colleagues have suggested that the only "real lawyers" are litigators, because litigation is the one task that, professionally speaking, only attorneys can officially perform. An attorney is an officer of the court with all the responsibilities that attach to that status. Litigation demands that lawyers employ all of their specialized methods of thinking: rules, risk, and conflict. Courts and arbitrators operate under a distinct and intricate set of rules and a client's case may be won or lost on detailed application of those rules. Litigators are constantly working under rules of conduct, rules of evidence, rules of procedure, and local rules required by the specific jurisdiction or judge.

And within these rules the lawyers must analyze risk. Courtroom procedure is always a matter of risking a client's money, pride, reputation, rights, and even freedom. My legal mentor used to tell clients that "if you have a dead-bang winner of a case, that gives you a 20 percent chance of losing." Many aspects of litigation fall beyond the control of even the best litigator. Whoever makes the final decision in court, whether judge or jury, may be swayed by the behavior of an attorney or party to the case, may be influenced by attitudes internalized long before the trial, may have tuned out during introduction of important evidence, or may simply be capricious. In addition, litigation usually involves not only the risk of winning or losing, but the additional risk of amount of money awarded or fined, or number of months/years of freedom a court takes from a criminal defendant. A famous recent study even established that prisoners were much more likely to receive a lenient parole award if their hearings were scheduled at the beginning of the day or immediately after the judge's lunch break, and that other factors like ethnicity or the severity of the underlying crime did not make a similar difference.[8] Risk management is difficult but necessary in litigation of all sorts. Many lawyers even build their client's risk into their own lives by operating on a contingent fee, where the attorney's team only receives payment if they win the litigation on behalf of the client, and the team receives a percentage of the award.

The essence of litigation is conflict, created by two adverse parties, each with a chance to argue for their cause and an equal opportunity to attack the positions advocated by the other. A judge, jury, or arbitrator will

---

8. Ben Bryant, *Judges are more lenient after taking a break, study finds*, THE GUARDIAN, Apr. 11, 2011. The article quotes Jonathan Levav, Columbia University professor and co-author of the study as writing, "You are anywhere between two and six times as likely to be released if you're one of the first three prisoners considered versus the last three prisoners considered." According to the study, "The only other variables that influenced a judge's ruling were the number of times a prisoner had been to jail and the presence of a rehabilitation programme. Other factors, such as the severity of the prisoner's crime, prison time, sex and ethnicity tended not to exert an effect on the rulings."

make decisions on whether the parties are properly making their cases within the rules, which of the parties wins, and what will be the effects of victory. Except in small-claims court where parties frequently act without lawyers, the stakes in court are relatively high because litigation costs are high, often placing courtroom justice beyond the reach of the average person. Even lawyers who never step into a courtroom need to be thinking about litigation conflict, risk, rules, and consequences. For transactional lawyers, contracts may fall apart and the relationships underlying those contracts may end in a courtroom. For patent preparation lawyers, their patent applications may be challenged in court or administrative proceedings. So even preparation of basic documents must be drafted and negotiated with possible litigation in mind. Litigation, the formalized, ritualized version of modern combat to solve problems, forces lawyers to be considering conflict strategy at every stage of their jobs.

A lawyer's mental processes should be marked by precision. Legal writing can seem convoluted and absurd to some people. The word "legalese," defined by the *Cambridge Dictionary* as "language used by lawyers and in legal documents that is difficult for ordinary people to understand," was coined to capture the apparently unnatural language used by lawyers. When written correctly, legalese is a sign of precise word choice rather than an intent to convolute the text. Particularly in contracts, an attorney is not seeking to make the language flow naturally or hold a reader's attention, but to express exactly what obligations and benefits the parties accept. Contract lawyers must try to anticipate conflicts that may arise years later, and thereby try to either minimize those conflicts or encourage the conflicts to be resolved in their clients' favor. The bigger the deal, the more important precision will be to its memorialization. Billion-dollar corporate acquisitions often involve teams of lawyers on both sides. These teams fight to hold an advantage for their clients, but also to pin down the most precise expressions of how the clients intend to benefit from the transaction. Are particular assets involved in the deal? If so, the lawyers list them all where possible and otherwise group them accurately. What happens in transitions between the time the deal is agreed and the time the deal can be practically completed? There is a transition services agreement for that. When you buy a house, your lawyers have very specific terminology and accurate descriptions honed over more than a century of similar transactions, so you can be sure to receive the rights you expect to live in the home and enjoy the land. Precision is everything.

Contract drafting lawyers are often seeking a precise set of benefits for their clients, even by leaving options open and by refusing to specifically define their terms. In other words, when ambiguity benefits a lawyer's client, then indefinite language is the best way to cement that benefit into a contract. So a lawyer's precise understanding of the contracting parties' relationship may lead to the conclusion that drafting imprecise language into an agreement is the best strategy for the client's benefit. So, in a good attorney's hands, sometimes vagueness and ambiguity can deliver the most precise client benefits.[9]

Litigators also use a specialized language built for precision. Arguments in court reflect the intricate rules restricting speech, and they reference other cases where similar conflicts have arisen. Proving a case in court often requires that exact language be stated in exactly the right way. For example, a prosecutor may need to prove that the alleged crime occurred within the county where the trial was taking place, so she would need a witness to the crime to identify the location, including the county. Otherwise, if the prosecution rested its case without introducing this evidence, the defendant may request the court toss the case for failing to establish jurisdiction. In a civil contract case, the plaintiff's counsel would need to enter the subject contract into evidence and to establish that an appropriate representative of each party executed the contract, or that the parties otherwise agreed to the contract's terms. If the exact point of fact is not introduced, then the plaintiff cannot succeed in his case. Precise language is crucial in any litigation.

Litigators are taught not to ask a witness questions to which the litigators don't already know the answer. Introduction of evidence at trial must support the lawyer's story of the case as efficiently as possible, so the judge or jury can understand the case in the manner the lawyer intended. Exact language matters so much in courtroom trials that lawyers train their witnesses to only answer the question asked, and to go no further. Offering extra commentary can muddy an otherwise clear case, or may grant the opposition an opening to press their own advantage. Precise presentation of evidence is the cornerstone of a successful case.

Law in the United States is based not only on statutes and rules, but the influence and interpretations of earlier court decisions. So litigators

9. This type of thinking is one of many reasons that the general public becomes frustrated with attorneys.

are specific in finding authority for their positions and tying that authority to behavior in the case at hand. Much U.S. law is based on precedent. A court generally looks to see if there is another case where a similar question was raised, and then the judge uses that case to support a decision. The litigants look for precedential cases that most closely align with the positions they are advocating, and then try to convince the judge that their historic cases are better precedent in this matter than the other side's historic cases. This entire exercise is based on precise advocative thinking and exacting research to find the best and most persuasive matches.

To see the need for citation precision, we need only to examine a famous misuse of generative AI in legal cases. In the federal case of *Mata v. Avianca, Inc.*,[10] lawyers for the plaintiff filed a motion response written entirely by generative AI, but these lawyers did not check all of the language and citations included in the document. To quote the judge when sanctioning the responsible attorneys, these attorneys "abandoned their responsibilities when they submitted non-existent judicial opinions with fake quotes and citations created by the artificial intelligence tool ChatGPT, then continued to stand by the fake opinions after judicial orders called their existence into question." When their opponents reviewed the filing, they found, "Although Plaintiff ostensibly cites to a variety of cases in opposition to this motion, the undersigned has been unable to locate most of the case law cited in Plaintiff's Affirmation in Opposition, and the few cases which the undersigned has been able to locate do not stand for the propositions for which they are cited."[11] The court conducted its own search for these "cases" cited in the response and, unable to find them, the judge asked the plaintiff's counsel to provide copies of the cases. Once again, rather than perform the required research, the plaintiff's counsel instead asked its generative AI model if the cases were real and to provide copies. The generative AI model produced a group of fictitious documents that were written in the mode of court filings and presented them as the underlying cited cases. The plaintiff's counsel presented these to the court with a signed affidavit claiming that the citations described real cases.

However, the original citations were false and ChatGPT simply made up publications that it claimed described the original false citations.

10. 678 F. Supp. 3d 443, No. 22-cv-1461 (PKC), 2023 WL 4114965 (S.D.N.Y. June 22, 2023).

11. *Id.*

Generative AI models do not have insights or make factual conclusions; generative AI models make predictions. Sometimes those predictions can present themselves as if they were correct factual conclusions (even when they are not), and, as generative AI models continue to improve, many more of these predictions will correctly answer queries.[12] But that does not mean that these predictions can be assumed to be correct statements of law, fact, and precedent.

In the *Mata* case, the court noted that while the snippets provided "have some traits that are superficially consistent with actual judicial decisions," severe and obvious deficiencies were clear to anyone reading the "cases" critically. Describing just one of these alleged supporting cases, the court noted that "[i]ts legal analysis is gibberish" and "quotation marks are often unpaired." The judge states that the "decision abruptly ends without a conclusion." The citations within the case do not exist or, when they do exist, do not contain the quotes attributed to them. In other words, any lawyer who viewed this text even in passing would see immediate red flags that the case was not correct, or was not a real case at all. Lawyers are required to present accurate information to the court, and for not doing so, the lawyers presenting AI-generated gibberish to the court were sanctioned and fined. Other lawyers who have also submitted false AI-generated citations to a court have suffered worse, with one young lawyer being both fired and suspended from the practice of law for such a failure.

Lawyers have a professional obligation of competence under their ABA Model Rules of Professional Conduct, some version of which has been enacted by the bar of every state, and this obligation extends to competence in use of technology in the practice of law. The commentary to the Model Rules explicitly state that "To maintain the requisite knowledge and skill, a lawyer should keep abreast of changes in the law and its practice, including the benefits and risks associated with relevant technology, engage in continuing study and education and comply with all continuing legal education requirements to which the lawyer is subject."[13] So while lawyers must keep up with changes in technology, it is crucial that everything that a litigator submits to a court over his name be reviewed and

12. Reviewing a generative AI model's result can trigger an insight in the reader, but do not mistake the helpful text for an insight itself. The generative AI can only predict which word should come next to answer your query, it does not think or see anything.

13. ABA Model Rules of Professional Conduct, Rule 1.1, Comment 8.

confirmed to be truthful and properly supported. It wasn't the misuse of the new technology that led to sanctions in the *Mata* case, but the inaccuracies in citations and subsequent misleading of the court. A lawyer is required to be precise and accurate in official statements, and to make certain that any document is correct. To do less violates a lawyer's duty of competence.

Finally, when contemplating how lawyers think differently than other people, we should examine the attorney's role as a policymaker and student of policy. In other words, lawyers are trained to examine how society addresses problems and to either operate within those parameters or to try to change the rules. One of the most common definitions of policy is a deliberate system of guidelines to influence decisions and achieve rational outcomes. Sometimes polices result in formal laws, rules, and regulations—the natural area of legal study. In other cases, policies lead to informal actions or inactions that can be just as impactful as official laws. Lawyers are trained to understand how policy works and how it can be harnessed to assist clients. Since the United States' independence from England, more than half of all presidents, vice presidents, and members of Congress have come from a law background.[14] So U.S. lawyers have always been active in making policies and laws, not just advising on policies.

But non-policymaker attorneys are active in policy analysis. Some of this work can be described as lobbying. If an attorney representing the pharmaceutical industry can argue for a change in policy that helps her clients, then she may be talking to Senators or the leaders of the U.S. Food and Drug Administration to propose such changes. The same can be true at local levels of government or even within private organizations. Convincing a university to study an important social issue may lead to important societal and/or client benefits. Some of this work involves nothing more than strategic decision analysis. Lawyers are often asked to analyze policies and social or contractual situations and advise clients the most effective course of action. Advising clients on when to talk to the press, how to present an issue, and whether to spend on a new product or course of action, all of these are standard work for any lawyer. Lawyers who work in-house

---

14. Nick Robinson, *Declining Dominance: Lawyers in the U.S. Congress*, Harvard Law School, Center on the Legal Profession, https://clp.law.harvard.edu/knowledge-hub/magazine/issues/lawyers-in-politics/declining-dominance/.

at organizations are generally viewed as important strategic members of the management team who can analyze internal risk-based policies.

Policy orientation can be especially important when thinking about new technologies. Lawyers stand outside of tech creation and critically analyze how the technology affects business, government, and society. The attorneys' policy inclination leads them to examine whether current laws or regulation will affect use of a new technology, or whether the technology's impact will need to be addressed by new rules. Lawyers examine the risks associated with the new technology and whether those risks fall evenly across society. They also review the advantages of the new technology and how clients can draw the most benefits from those advantages. A general feeling for policy allows lawyers to see a new technology within the context of current and future society, analyzing how that technology affects people and what, if anything, should be done to maximize benefits and minimize problems for the rest of us.

When a lawyer examines the introduction of a new technology into our world, the lawyer's view is filtered through specialized training in risk, rules, conflict, and policy, with a focus on precision thinking. "Thinking like a lawyer" means applying all of these filters to develop an understanding of how novel technologies will impact society in the immediate term, and in the long run. Stanford futurist Roy Amara is known for saying that the impact of a new technology is often overestimated in the short run and underestimated in the long run.[15] Lawyers need to understand the new technology as it affects the world—their clients, their communities, their practices—right now and going forward. They apply their training filters and their precision thinking to see the technology in context, not as a novelty, but as a tool. How technology works is not significant to most lawyers unless it affects how technology creates risks, breaks rules, leads to conflicts, or affects policy.

Looking at the world this way is neither good nor bad, but it can be useful in many situations. Thinking like a lawyer about artificial intelligence can lead to a better understanding of how our society should tackle the technology, and assess whether we already have the necessary policies in place to address personal risks that the technology creates.

---

15. Doc Searls, The Intention Economy: When Customers Take Charge 25 (Harvard Business Press 2012).

CHAPTER 3

Lawyers Should Think of AI in Functional Categories

It is easy to become overwhelmed when asked to contemplate AI. Generative AI has burst into everyone's attention, but there is a different mode of AI in our word processing programs, our mobile shopping applications, and our online chess tutorials. A distinct type of AI searches for tumors in medical tests and another type operates drones that target warriors from a mile in the sky. Non-generative AI is the key to facial recognition systems and fingerprint identifiers. Multiple disparate applications are too much to consolidate into one word.

And yet, many lawyers are doing so. As described in chapter 2, the "hallucination" problems of generative AI, most famously illustrated by the lawyers' mistakes in the New York jurisdiction *Avianca Airlines case*,[1] led to what seems like a panicked response from many U.S. courts. Lawyers in New York relied on generative AI to write a court filing for them, and the filing included fabricated citations. When called on the falsifications, the lawyers doubled down and defended the indefensible (primarily because they asked the AI if it was lying and the AI doubled down on its predictive generations). The New York judge disciplined and fined these lawyers under Rule 11 of the Federal Rules of Civil Procedure for trying to pass off fictional case citations as real historical facts.

1. *Mata* v. *Avianca, Inc.*, No. 1:2022cv01461—Document 54 (S.D.N.Y. 2023).

Relatively quickly, judges throughout the United States were issuing orders restricting and building requirements around the use of AI in their courtrooms. But these orders often simply proscribed the use of any artificial intelligence models, not just generative AI. One of the earliest orders was issued by Judge Michael M. Baylson of the U.S. District Court for the Eastern District of Pennsylvania, who wrote, "If any attorney for a party, or a *pro se* party, has used Artificial Intelligence ('AI') in the preparation of any complaint, answer, motion, brief, or other paper filed with the Court, and assigned to Judge Michael M. Baylson, they **MUST**, in a clear and plain factual statement, disclose that AI has been used in any way in the preparation of the filing and **CERTIFY** that each and every citation to the law, or the record in the paper, has been verified as accurate."[2] (Emphasis in original) Knowing that non-generative AI models are included in dozens of legal tools like Microsoft Word and in online research services like Westlaw and Lexis/Nexis, it would be non-sensical for any lawyer in Judge Baylson's court to be held responsible for failing to disclose all of the AI that was tapped in writing a submitted document. The judge has reacted to generative AI hallucination by demanding disclosure involving all forms of AI, which would seem to include the most benign AI tools, like spelling and grammar checking programs.

The disparate functionalities captured in the simple term "AI" make the judge's intent unclear on the day it was written. But this is not the only significant problem of overinclusive definitions (or failure to define in many cases). Even when limited to generative AI models, the tools available to lawyers at this moment may be substantially different next year than the tools currently at our disposal. The concern around these reactive court orders are even more substantial when one considers the rapid adoption, evolution, and expanding applications for all forms of AI, even for generative AI.

> An excellent article published by Duke University School of Law notes:
>
> Most search engines and word-processing systems, for example, will soon embed the use of large language models (LLMs), a type of GenAI trained on massive data sets to recognize, translate,

2. *Standing Order Re: Artificial Intelligence ("AI") in Cases Assigned to Judge Baylson* (E.D. Pa. 2023), https://www.paed.uscourts.gov/documents/standord/Standing%20Order%20Re%20Artificial%20Intelligence%206.6.pdf.

> predict, or generate text in a human-like fashion. Rules of civil procedure should be technology-neutral and should not have to be revised with the introduction of each new technological development. No one can predict what the legal technology environment will look like two years from now, but the use of GenAI will almost surely be ubiquitous.
>
> The legal profession is already sufficiently risk averse and technologically backward. These orders will impede innovation and chill the use of technology that could not only enable unrepresented parties to access the justice system, but also reduce the time and cost for those who can afford representation.[3]

The generative AI revolution has already caused a cascading failure effect on the legal profession. By misusing the tools and by not following the most basic rule of competence–the rule that an attorney must understand and be able to defend every word in court documents filed under their name—some lawyers have inflamed the judiciary (and plenty of legal organizations) into overreaction. A significant number of courts and law firms simply banned the use of the technology in their fields of control, while others overreached by setting unrealistic rules which did not carefully define the core issues of concern.

Paradigm shifting new technology spurs overreaction in the legal profession. When the fax machine was introduced, many law firms refused to participate because of concerns over client confidentiality. The same was true when email first became an essential tool of business. Searchable online legal databases were expected to kill the research abilities of young lawyers, rendering them incapable of finding and using appropriate citations. Cloud computing was resisted for security and confidentiality reasons. Some lawyers are certainly early adopters of the newest tech. But a large and influential set of legal leaders tends to either resist major technological changes or overreact to the perceive threats these advances hold for the legal profession.

Lawyers cannot bury their heads in the sand and hide from new technologies; they cannot avoid an understanding of AI or other technology being used by firms and clients. As described in chapter 2, where

3. Maura R. Grossman, Paul W. Grimm & Daniel G. Brown, *Is Disclosure and Certification of the Use of Generative AI Really Necessary?*, 107(2) JUDICATURE 71–72.

technology affects clients or practice, a lawyer is bound to learning the necessary information about that tech. A lawyer is responsible under the rules for providing competent representation to clients. "To maintain the requisite knowledge and skill, a lawyer should keep abreast of changes in the law and its practice, including the benefits and risks associated with relevant technology."[4] This applies not just to the technology that a lawyer chooses to use in practice, but also to the technology that clients may be using or that may create an issue of law that ensnares a client.

While attorneys have carved out an insulated sphere of protection and action within their own courts and practices, lawyers must address technology sweeping over society on its own terms. Whether or not lawyers choose to apply or avoid generative AI tools in their own practices, their clients, colleagues, and children will be driving further into generating images, text, and code, and those new developments will create legal issues to be resolved. Generative AI models are already deeply influential as the sources of regulation and case law, even if a plurality of lawyers have never entered a prompt into the model. And other forms of AI are also creating legal issues addressed by the courts, legislatures, regulators, and executive branches. Lawyers will need to address AI—generative and otherwise—as a subject of legal consternation and dispute.

So how should a lawyer be thinking about artificial intelligence? How does all of the language precision, conflict focus, policy orientation, and risk management discussed in chapter 2 combine when considering a rapidly evolving technology overtaking important aspects of our lives and practices?

I argue throughout this book that lawyers should lean toward the more practical aspects of their expertise and see AI as a series of tools defined by functionality. Because AI performs so many tasks in so many fields—and those tasks and fields will continue to grow at an alarming pace—it makes sense for lawyers to consider which AI functions can be grouped by legal impact. Obviously, different functions create different legal problems. Intellectual property issues are raised by generative AI, but not by most AI that optimizes systems. Constitutional issues are raised by AI that allows government agents to achieve biometric identification of people in peaceful crowds. Harmful bias is a significant legal and technical

4. Model Rules of Professional Conduct, Rule 1.1, Comment 8.

problem for AI that facilitates hiring decisions. Violation of privacy rights is a possible result of AI that reads brain electrical signals and translates thoughts into words or actions.

There are a few simple categorizations by functionality that seem natural to me and to the digital technology lawyers I work with. Each of the major categories is given its own chapter below. Each category triggers a different set of laws and legal concerns based on the way the AI models are currently used in our society. This means that eventually, as AI tools continue to evolve and find new applications, the categories in this book will divide and change. It could soon be true that as we find new uses for large language models (LLMs), we will divide the AI down into those generative products that create audiovisual material, those that write documents in the vernacular, and those that create computer code. However, I also expect that we will soon see tools combining all of these capabilities. We cannot predict the future, but this model of legal thinking about AI will remain valid because the risks invoked by practical functionality should be the core of legal thought about new technologies.

Categories may change, but the importance of practical observation remains. As noted in the previous chapter, attorneys focus their attention on risk, and risk follows function. The importance of AI to an attorney arises not from the type of programming used to create the model (the method that many technologists use to define AI), but from how that model operates in the real world and the legal issues the operations generate.[5] Building a framework on functionality and legal risk can help attorneys be more precise in their definitions and, therefore, be more effective in their legal analysis. More precise definitions based on functionality is also crucial to policymakers. Writing laws to manage a fast morphing technology is tricky. Definitional vagueness makes it harder to achieve policy goals without unnecessarily stepping into unanticipated minefields. Overregulation can squelch the growth of a technical tool and give other societies an edge in technological development. So starting with the essential

5. Please note that the method of creating AI models may birth its own set of legal risks. Most of the AI models we discuss in this book were created by feeding enormous data sets into an algorithm. Legal questions have already arisen about the biases in some of those data sets and whether the AI creators had the right to use the data sets to train and test the AI models. The Federal Trade Commission has already ordered the destruction of certain AI models due to irregularities in the training data sets. This is discussed in more detail later in the book.

practical questions, "What does this tool do and who does it affect?" can make both law and policy better.

There are two important operationally-based definitions for the world of machine learning, large language models and artificial intelligence that will not be the focus of chapters in this book: extractive AI and predictive AI. While both of these descriptors are important for both technical and functional purposes, they are both too broadly inclusive to be useful to the legal analysis recommended in this book. Both are terms that lawyers should understand, but these terms are not necessarily practical categorizations for grouping legal risk.

Extractive AI taps logic and search functions to pull a set or sets of information out of a database. Unlike generative AI, extractive AI does not predict what word or pixel comes next to provide an output to a query. Instead, the extractive AI runs a sophisticated review of a specific database and pulls items out of the database that are responsive to the query. Extractive AI is valuable to the legal profession, and dozens of extractive legal tools exist at this writing. Searching LEXIS or Westlaw databases is a well-known use of extractive AI, as basic Boolean search algorithms were replaced by more sophisticated tools trained on machine learning methods. The enormous databases of cases and legal analysis can be mined with questions beyond simple word or phrase searches, so that concepts may be extracted that could not have been gathered a decade ago.

Using extractive AI, a transactional lawyer may ask to find the variations of indemnity clauses contained in all of a company's vendor contracts within its procurement database. A litigator may feed hundreds of thousands of pages of depositions into the extractive model and request a list of references to certain relevant concepts or phrases. A corporate lawyer may use extractive AI to find all risk-based statements regarding investments in the SEC filings of a client company—or the client's rival companies. In all these cases, the sophisticated conceptual evaluation of an extractive AI is applied to the files at hand, and all the output provided in response to prompts will be found within the bounds of those documents. So lawyers should know that AI exists and more advanced versions are being developed that will extract information only from the items at hand. Many lawyers and their clients are using these tools. However, as a category of functionality, document and database extraction does not raise the kind of legal risks we describe in the categories below. Document extraction is a relatively simple and benign use of AI. Load the documents into the AI model and ask the AI model questions about their contents. This exercise does not usually raise legal concerns or impinge on the rights of others, especially if the AI model user has the clear right to extract information from the chosen documents.

For a different reason, predictive AI is not awarded a chapter in this book. While extractive AI does not raise significant legal issues, predictive AI is a broad category that creates risks of many types. The name "predictive AI" describes too much of the underlying work that allows several AI functionalities to work. For example, generative AI is predictive, as it makes predictions about which incremental outputs build on previous outputs to respond to a prompt. Decisioning AI is predictive, and may be based on predictions that the person represented by a loan application will have the income and other ability to repay a loan, will live peaceably in the landlord's available apartment, or will be appropriate to interview for a cloud database team leader position. The term "predictive AI" is overinclusive for our purposes, as nearly every functional category described below involves some level of sorting, grouping, and predicting. The prediction attribute is crucial to today's useful AI, and it is wrapped into nearly all of the legally-significant groupings discussed here.

Some technologists define predictive AI more narrowly, only encompassing AI models that analyze information about the past to make predictions about the future.[6] This limitation also does not satisfy our needs because a prediction without context or task-specificity does not meet our requirement of a functional category. Using past to predict future only becomes a legal issue in the context of its application. Predicting future weather patterns may have some liability risks, but is essentially not a legal matter. Predicting the likely recidivism of current parolees has definite legal effect. So we acknowledge that prediction, however we choose to define it, is an important aspect of many types of functional AI, and therefore is an aspect of AI that lawyers should note. Given that prediction is such an inherent attribute to many risky types of AI, the more important question for lawyers is "Will the predictive aspects of a generative AI model or a strategizing AI model affect the legal risks involved in using

6. More broadly-defined predictive AI makes predictions based not just on past events, but can include current conditions, specific attributes of a person, place or system, or any other fact that may be useful.

those models?" We must look closer at legal impact and applicability to situations that are significant to our society.

What are those categories of AI functionality and legal impact in 2025? Of course, one category is generative AI. As noted, this is a tool that demonstrates many capabilities, from chatting with a company's customers to designing the code for new websites, and from writing legal briefs to creating art in the style of van Dyck. So we could expect this category to split sometime in the future as the legal issues arising from generative AI models may distinguish one set of functions from another, but for now, with generative AI still remaining a relatively early stage technology for most public functions, I am examining all of these functions in one category with a broad set of legal issues.

Because generative AI creates entirely new content when prompted by people, and because the methodology of creation is almost entirely machine-driven, then generative AI challenges our established systems of intellectual property protection. Copyright and patent, the centuries-old systems of content protection, were not written to meet this moment. The framers of the U.S. Constitution believed that allowing authors to protect their writings would "promote the progress of science and useful arts," by allowing the authors to financially benefit from their works, which would simultaneously benefit the general public by encouraging those authors to share their work with the rest of us. So the ability to protect a writing, painting, or even lines of computer code by encouraging productive distribution while squelching copies was always an exercise in promoting original thoughts in publication. The law will protect an original work created by a human because the policy behind this law encourages people to create, to secure their creations into a tangible medium, and to expose the creation to the world.

But how much of the act of "creation" does the human need to participate in? We have awarded copyrights to corporations and other entities, although the initial creator has always been assumed to be human. Carefully and creatively prompting an AI to generate specific output could be considered an art form, and if it is, then is work created through that prompt an artistic work of the human who imagined it? Not according to the U.S. Copyright Office, who ruled, "In the Office's view, it is well-established that copyright can protect only material that is the product of human creativity. Most fundamentally, the term 'author,' which is used in both the Constitution and the Copyright Act, excludes non-humans. . . .

applicants have a duty to disclose the inclusion of AI-generated content in a work submitted for registration and to provide a brief explanation of the human author's contributions to the work."[7] The Copyright Office will grant copyrights in the aspects of works created by humans, but not those created by AI models. Even for works that contain both human-generated and machine-generated content, only the human content will be protectable by the U.S. government copyright regime.

The U.S. patent office takes the same position. In 2020, the patent office ruled that only "natural humans" can apply for recognition as inventors of patentable technology, finding that the plain language of the patent laws passed by Congress and interpreted by the courts supports this view. Early petitioners for AI inventorship cite a number of thoughtful policy reasons, including the fact that crediting AI as an inventor would encourage people to create and publicize AI inventions—to the benefit of everyone. However, the years of legal precedent overwhelmed this logic. A number of other intellectual property issues also arise with the introduction of generative AI, but we will cover those later in a dedicated chapter.

Contrast the generative AI legal issues with those raised by identifying AI, a technology that capitalizes on AI's advantages in sorting, grouping, and excluding vast amounts of data to make biometric identifications of humans. Identifying AI is the tech used by the Federal Bureau of Investigation (FBI) and other law enforcement agencies to compare the loop and whorl patterns on a person's fingerprints with those in the gigantic fingerprint databases to allow a sixteen-point match to the suspect's fingerprint found at the crime scene. Identifying AI may also match people by any number of unique persistent identifiers, such as face geometry, vocal inflections, retinal patterns, and even gait. This technology has raised controversial issues and spawned a series of state and local laws limiting or prohibiting its use. Most famously, Illinois's Biometric Information Privacy Act (BIPA) has been the basis of hundreds of millions of dollars of penalties for applying identifying AI without the express written permission of data subjects. Other states and cities have prohibitions on the use of facial recognition technology by government or businesses. None of these biometric identifications could be practical without identifying AI

---

7. *Copyright Registration Guidance: Works Containing Material Generated by Artificial Intelligence*, A Rule by the Copyright Office, Library of Congress, Mar. 16, 2023.

which can both analyze the key features provided by a biometric measurement and then compare those key features across millions of other samples to find prospective matches.

U.S. Constitutional issues arise from investigations including identifying AI. The First Amendment to the Constitution allows for freedom of association, and the U.S. Supreme Court has held that anonymity of association is also protected by the First Amendment. So when the government, through law enforcement or otherwise, captures the pictures of a peaceful crowd of political protesters and then runs those pictures through a facial recognition AI model to find the names and lives of those protesters, the government may be violating the First Amendment rights of those citizens. Identifying AI provides powers to the government that the government would never have otherwise, destroying the "security by obscurity" of every citizen who shows their face out in the street.

The last example we will cover in this chapter is AI that operates machines that affect the physical world, or physical AI. This AI can operate machines on an assembly line, direct driverless taxis now running in the streets of San Francisco, or manage spy drones/attack drones in a war zone. In each case, the physical AI is responsible for navigating real-world spaces, often with heavy and/or dangerous equipment, in the vicinity of humans, animals, and other unpredictable and fragile obstacles. What could go wrong?

Physical AI relates with the world by accepting and processing information taken in by sensors and then responding or interacting with the world through actuators. The AI is designed to interpret the sensory input and translate the response into appropriate action. Clearly, this scheme leaves at least three levels of potential failure in the system. Sensors could either fail or be blocked; for example, early self-driving vehicles did not operate correctly in the rain. The AI could fail at interpreting the sensory input or could fail in choosing a corresponding action. Those failures may reflect mistaken analysis, poor training, or even a completely novel situation presented to the AI. Finally, the actual physical machine could behave improperly, due to need for repair or other physical impairment. Flat tires happen to nearly all of us.

Futurist Ray Kurzweil's predicted nanobots are not yet coursing through our bloodstreams and hooking our brains into the global consciousness, but we still have legal risks in the present state of physical AI implementation. The fact that currently-constituted physical AI operates in a world of gravity and other laws of physics adds frightening elements of complexity and consequence to its management. Where a generative AI mistake may result in a copyright violation, a physical AI mistake may result in a busload of dead citizens. We currently have criminal laws that address human foibles, follies, failures, and malice in the physical world. How will those criminal laws—and associated civil damage-calculation laws—apply to an AI-driven semi-truck or a warehouse robot that bashes its human co-worker? How will such non-human operators be insured and will their insurance cover those of us affected by their mistakes? Some physical accidents happen despite excellent behavior by all parties involved. Who would defend the AI driver in such a circumstance? These are questions we address below.

Maintaining a realistic understanding of AI starts with dividing the topic down into manageable and logical categories. Lawyers' practical and risk-based focus on real-world problems should lead them to look at AI as an operational problem, grouped into functions that present a unique set of legal issues. This book suggests how several of these functional categories can be examined and regulated. The categories may change as the applications and risks evolve, but the system of dividing and understanding should continue to be valid.

CHAPTER 4

Decision-Making AI

At the time of this writing, the United States imprisons more people than any other country in the world.[1] According to the Bureau of Justice Statistics, the primary statistical arm of the U.S. Department of Justice, more than 5.4 million people in the United States were under correctional supervision at the end of 2022, meaning about 1 in 48 Americans was in jail, prison, or probation (down from 1 in 35 in 2012).[2] Addressing this enormous population and the resources needed to manage it has led U.S. authorities to employ AI in the form of algorithmic decision-making. From predictive policing tools to pretrial risk assessments, sentencing algorithms, prison management software, and parole decision predictors, the policing, sentencing, and incarceration industry is one of the earliest adaptors of AI decision-making.

Many policymakers have been tempted to apply AI tools to the famously difficult task of predicting criminal recidivism. As noted in a recent law review note, "It is argued that regardless of how much rigorous legal training a judge has, it is impossible to strip a practice like this of implicit bias. This is coupled with the fact that humans do not seem to be particularly good at predicting the future behavior of other humans, including recidivism."[3] The note goes on to cite an example where sitting

1. *Countries with the largest number of prisoners as of December 2023*, Statista.

2. This includes 1 in 29 American males and 1 in 141 American females. Emily D. Buehler & Rich Kluckow, *Correctional Populations in the United States, 2022 – Statistical Tables*, Summary, NCJ 308699, Bureau of Justice Statistics, May 2024.

3. Kieran Newcomb, *The Place of Artificial Intelligence in Sentencing Decisions*, University of New Hampshire, Inquiry Journal, Spring 2024.

judges made correct recidivism predictions in 66.7 percent of cases while the Correctional Offender Management Profiling for Alternative Sanctions (COMPAS) program described below was accurate in 65 percent of cases. In the same law review note, the authors argue, "Neither of these results inspires a lot of confidence, given that random guessing should yield about 50% accuracy."[4] But we assume that the AI tools can improve over time, whereas human implicit bias remains consistent.

I have already established that certain machine learning programs make useful predictions. These predictions can suggest weather patterns or likely chess moves, but they can also underly decisions that affect people's lives. For example, sentencing judges may benefit from an accurate prediction of how likely a convicted felon would be to commit a subsequent crime. If reducing future crime is one of the purposes behind sentencing a criminal to more years in prison, then a judge's sentencing decision may be grounded in science if influenced by AI using scientific prognostication methods. On the other hand, this type of process may create the illusory scientific cover known as the CSI effect—where decision-makers grant unwarranted weight to facts and determinations that seem to arise from a scientific system. In any case, a person's life is deeply affected by a sentencing decision, so the reasoning for a ten-year prison sentence rather than a four-year prison sentence should be as thoughtful and policy-minded as possible.

A handful of states use risk-assessing AI to help judges determine sentences in criminal cases.[5] In 2012, the state of Wisconsin has used a predictive program called COMPAS, which was developed by a private company Northpointe (now called Equivant). Use of the COMPAS tool was unanimously affirmed by the Wisconsin Supreme Court in 2016,[6] and seems to still be in use at the time of this writing. While noting that the risk to reoffend is the cornerstone of effective criminal supervision, the Wisconsin Department of Corrections website explains the COMPAS AI tool by writing, "This actuarial risk assessment system contains offender

---

4. *Id.*

5. According to *ProPublica*, Arizona, Colorado, Delaware, Kentucky, Louisiana, Oklahoma, Virginia, Washington, and Wisconsin all use risk-assessing algorithms for criminal sentencing. *See* Julia Angwin, Lauren Kirchner, Jeff Larson & Surya Mattu, *Machine Bias: There's software used across the country to predict future criminals. And it's biased against blacks*, ProPublica, May 23, 2016.

6. *Loomis v. Wisconsin*, 881 N.W.2d 749 (Wis. 2016), cert. denied, 137 S. Ct. 2290 (2017).

information specifically designed to determine their risk and needs and inform dynamic case plans that will guide the offender throughout his or her lifecycle in the criminal justice system." To guide the offender, and the judge sentencing the offender, COMPAS predicts the likelihood of recidivism, provides a numerical risk score for repeat offenses, and suggests the term of offender incarceration and the level of supervision the offender will need. A local Madison newspaper explained how COMPAS makes predictions:

> When most defendants are booked in jail, they respond to a COMPAS questionnaire. Their answers are fed into the COMPAS software to generate several scores including predictions of 'Risk of Recidivism' and 'Risk of Violent Recidivism.' The risk scores are displayed in a bar chart, with three bars that represent pretrial recidivism risk, general recidivism risk and violent recidivism risk. Each bar indicates a defendant's level of risk on a scale of one to ten. Judges are given these scores and use them in determining how to sentence a defendant.[7]

COMPAS delivers data-driven recommendations to accompany sentencing reports provided by law enforcement to judges making sentencing decisions. In Wisconsin, these reports follow a convicted person through the penal system and may impact places of imprisonment and availability of parole.

Factors weighed by criminal justice decision-making AI models can vary depending on both the model used and the jurisdiction applying the model. The Wisconsin version of COMPAS can include factors such as homelessness to predict future crime. According to *The New York Times*, "Utah employs a method that weighs what state officials call the 'Big Four' factors that they say most correlate with recidivism: history of behavior that harms others, antisocial personality patterns, attitudes and beliefs that favor crime, and 'association with pro-criminal peers.'"[8] The Offender Assessment System (OASys) AI algorithm introduced into the U.K. criminal justice system in 2001 is used for a variety of purposes including

7. Katelyn Ferral, *Wisconsin Supreme Court allows state to continue using computer program to assist in sentencing*, The Capital Times, Jul. 13, 2016.

8. Mitch Smith, *In Wisconsin, a Backlash Against Using Data to Foretell Defendants' Futures*, N. Y. Times, Jun. 22, 2016.

granting bail, assigning incarceration options, and immigration decisions, makes judgements on some immutable factors, such as criminal history and age, and some variable factors, such as accommodation, employability, relationships, lifestyle, drug abuse, alcohol misuse, thinking and behavior, and attitudes. OASys assigns different values to these factors.[9] There seems to be no consensus on what factors must be examined to predict criminal recidivism and how much weight should be applied to each factor input into the predictive AI model.

Northpointe maintains the methodology and functionality of the COMPAS algorithms as a trade secret, and will not make this information available to convicted offenders to analyze or dispute. This is the essence of the AI black box discussed in this book—a mysterious machine where inputs are entered and a recommendation or score is spit out with no explanation of how the score was calculated. In this case, the numerical score predicts a convicted person's likelihood of committing crimes in the future. So the AI tools are recommending that courts take away years of a person's life based on a secret computerized evaluation of the person's history and answers to post-arrest questions. *ProPublica's* 2016 investigation of COMPAS found racial and gender disparities in the AI model's sentencing recommendations that demonstrated bias against black offenders.[10] In an analysis of over 10,000 criminal defendants in Broward County, Florida, *ProPublica* found that COMPAS predicted recidivism rates correctly only 61 percent of the time and predicted violent recidivism correctly 20 percent of the time. The article concluded that COMPAS was biased because it performed worse on false positive rates for black individuals when compared to white individuals.

AI decision-making models have been used in institutions for decades. In the 1920s, Chicago School sociologists suggested using statistical regression analysis for Illinois parole board decisions.[11] Criminologist Richard Berk wrote that the Illinois parole board started to use algorithms to predict recidivism as early as the 1920s. The calculations and inputs were simpler then than now. Berk said, "In the '20s, parole boards were

9. Melissa Hamilton & Pamela Ugwudike, *A 'black box' AI system has been influencing criminal justice decisions for over two decades – it's time to open it up*, The Conversation, Jul. 26, 2023.

10. Julia Angwin *supra* note 5.

11. Rodrigo Ochigame, *The Long History of Algorithmic Fairness: Fair algorithms from the seventeenth century to the present*, Phenomenal World, Jan. 30, 2020.

worried about what parole boards are worried about today: If I release somebody, are they going to commit a horrible act?"[12] The movement to use algorithms and actuarial methods in the criminal justice system spread over U.S. states in coming decades, until it could be argued that the U.S. criminal justice system reflected insurance-like actuarial theories to manage society by removing problematic members. "Although the relationship between actuarialism and mass incarceration is complex, it is worth noting that the progressive adoption of actuarial methods coincides with the dramatic increase of the U.S. prison population since the 1980s and with the penological shift towards targeted interventions of crime control and risk management, away from midcentury policies of welfare provision."[13] The more that our society viewed convicted felon sentencing as a numbers game that could be scientifically calculated using ever-more-complicated algorithms, the easier it was to cast an individual convicted person's fate to the machines and the math. Some people believe that justice can be reduceable to numbers, without recognizing the complexity of what justice may be to different people within our society.[14]

AI decisioning tools are designed to take subjective decisions that affect a person's life, like the length of her prison sentence, her qualifications for a new job, or whether she is suitable to rent a particular apartment, and to make the answers seem objective by waving the mysterious wand of science-based predictions over them. The AI predictive model evaluates factual points from a person's life, runs those facts through the algorithm, then predicts whether that person is fit for the benefit applied for, or instead deserves a punishment, depending on the circumstance. The situation is not only dehumanizing, but, in the name of efficiency, it objectifies a deeply subjective process—the evaluation of a person's past life and what that life may say about how the person will live in the future. I shouldn't need to write this next phrase, but our futures are not

12. Richard A. Berk & Justin Bleich, *Statistical Procedures for Forecasting Criminal Behavior: A Comparative Assessment*, 12(3) Criminology & Public Policy 513–44 (2013).

13. Rodrigo Ochigame *supra* note 11, citing, David Garland, The Culture of Control: Crime and Social Order in Contemporary Society (Chicago: University of Chicago Press, 2001).

14. Using algorithms, AI models or other "scientific methods" also takes the psychic toll of responsibility off of individual human decision-makers. This effect may be helpful to longevity of judges and parole boards, but its effect on justice for affected individuals is debatable.

predetermined. If tomorrow comes for us, it will offer an opportunity to change our lives in small and large ways. Advocating for the continued use of human decision-makers in courtroom evaluations, U.S. Supreme Court Chief Justice Roberts wrote, "Judges, for example, measure the sincerity of a defendant's allocution at sentencing. Nuance matters: Much can turn on a shaking hand, a quivering voice, a change of inflection, a bead of sweat, a moment's hesitation, a fleeting break in eye contact."[15] While our increasingly-burdened institutions must find methods of making consistent and effective lending, leasing, and sentencing decisions across the board to allow successful functioning, the tendency to reduce these decisions to an unquestioned and unquestionable "scientific" and "objective" prediction can ignore the basic truth about humans—we are all capable of change, and therefor unpredictable. Like application of science fiction pre-cognition factors to predict crime, AI-based predictive decision-making takes away our life opportunities on a faulty assumption of scientific certitude.

Risk of bias is a clear issue for any decisions using a person's history and characteristics to award a significant benefit or impose a significant punishment. For example, when it comes to using AI models for bail decisions, "There is a very real danger that these tools and the appeal of 'objective risk scores' will silently codify racial disparities in bail determinations under a veneer of scientific rigor," says Scott Levy, director of the Fundamental Fairness Project at the Bronx Defenders, a legal aid organization. "It is essential that appropriate oversight and transparency mechanisms are in place."[16] There are a number of ways that AI decision-making models can cement bias into their systems. They might reflect the bias of the humans who are training and shaping the AI model. For example, developers of a machine learning tool may choose—knowingly or otherwise—to use databases that only contain information about people of one ethnic origin, or they primarily exclude information about people from another ethnic origin. Alternatively, AI decision-making models might be fed a bias that is contained in the training and testing data even if the developing humans try to exclude such bias. In this case, a conscientious

15. Chief Justice John G. Roberts, Jr., *2023 Year-End Report on the Federal Judiciary*, Supremecourt.gov.

16. Jason Tashea, *Risk-assessment algorithms challenged in bail, sentencing and parole decisions*, ABA Journal, Mar. 1, 2017.

AI model developer could attempt to include all sorts of datasets into the training material, but the sets themselves could lead to a bias in favor of men over women.

The AI model might receive unbiased training data but could still draw biased conclusions from that data. For example, the AI model may determine that people who live in a specific neighborhood or set of city streets are less likely to pay their bills than people in another neighborhood. While this conclusion may be rational, it could have arisen from historical and illegal discrimination of the people in the target neighborhood, so that the machine learning program is perpetuating unlawful bias. As observed in the *MIT Technology Review*, "[M]achine-learning algorithms use statistics to find patterns in data. So if you feed it historical crime data, it will pick out the patterns associated with crime. But those patterns are statistical *correlations*—nowhere near the same as *causations*. If an algorithm found, for example, that low income was correlated with high recidivism, it would leave you none the wiser about whether low income actually caused crime. But this is precisely what risk assessment tools do: they turn correlative insights into causal scoring mechanisms."[17] (Emphasis in original)

The authors of this article observe that now communities who have traditionally been targeted by police will receive higher AI predictive model scores because they have more arrests and convictions in the past. They write, "As a result, the algorithm could amplify and perpetuate embedded biases and generate even more bias-tainted data to feed a vicious cycle. Because most risk assessment algorithms are proprietary, it's also impossible to interrogate their decisions or hold them accountable."[18] Allowing bias to build this way is an insidious problem that can tie down minority communities with the bonds of past discrimination.

A different, but excellent example of this last category of biased AI involves a hiring system developed—and subsequently dropped—by Amazon several years ago. Since 2014, Amazon's machine learning specialists built a tool to help sort resumes to hire for open positions. The new tool

---

17. Karen Hao, *AI is sending people to jail—and getting it wrong: Using historical data to train risk assessment tools could mean that machines are copying the mistakes of the past*, MIT Tech. Rev., Jan. 21, 2019.

18. *Id.*

used AI decisioning to give job candidates a very Amazon-on-brand score on a 1–5 scale. According to a Reuters article, "But by 2015, the company realized its new system was not rating candidates for software developer jobs and other technical posts in a gender-neutral way."[19] Most of the resumes fed into the AI model came from men, who dominated the tech industry at the time. Reuters wrote, "In effect, Amazon's system taught itself that male candidates were preferable. It penalized resumes that included the word 'women's,' as in 'women's chess club captain.' And it downgraded graduates of two all-women's colleges, according to people familiar with the matter. They did not specify the names of the schools."[20] Amazon developers edited the AI model to neutralize these terms, but they could not be assured that the model wasn't finding other ways to promote men and degrade the applications of women, so they disbanded the development team and stopped using the tool in 2017. Gender bias was not the only issue presented by this AI model, as it also proposed unqualified applicants for jobs at Amazon. Policymakers can address algorithmic bias issues. For example, as discussed below, New York City now has a law aimed at reducing bias in hiring for companies using AI decision-making tools.

For technologists, bias is a systemic distortion that ruins a statistically significant result or, in the AI models we are currently examining, a prediction. For a lawyer, impermissible bias is an act or omission that unlawfully discriminates against people in a legally protected category, whether that bias arises from disparate treatment of the protected groups or from disparate impact of a policy that does not appear to follow from an intent to discriminate. Social scientists would regard explicit bias as the conscious choice to discriminate, while implicit bias is the subconscious discrimination that occurs without explicit intent. All three of these definitions will be important to understanding the cost of bias in an AI predictive system that assists in systemic decision-making. The following figure from the National Institute of Standards and Technology (NIST), a section of the U.S. Department of Commerce, demonstrates the complexities of biases at play in decision-making AI.[21]

---

19. Jeffrey Dastin, *Insight - Amazon scraps secret AI recruiting tool that showed bias against women*, Reuters, Oct. 11, 2018.

20. *Id.*

21. Reva Schwartz, Apostol Vassilev, Kristen Greene, Lori Perine, Andrew Burt & Patrick Hall, *Towards a Standard for Identifying and Managing Bias in Artificial Intelligence*, NIST Special

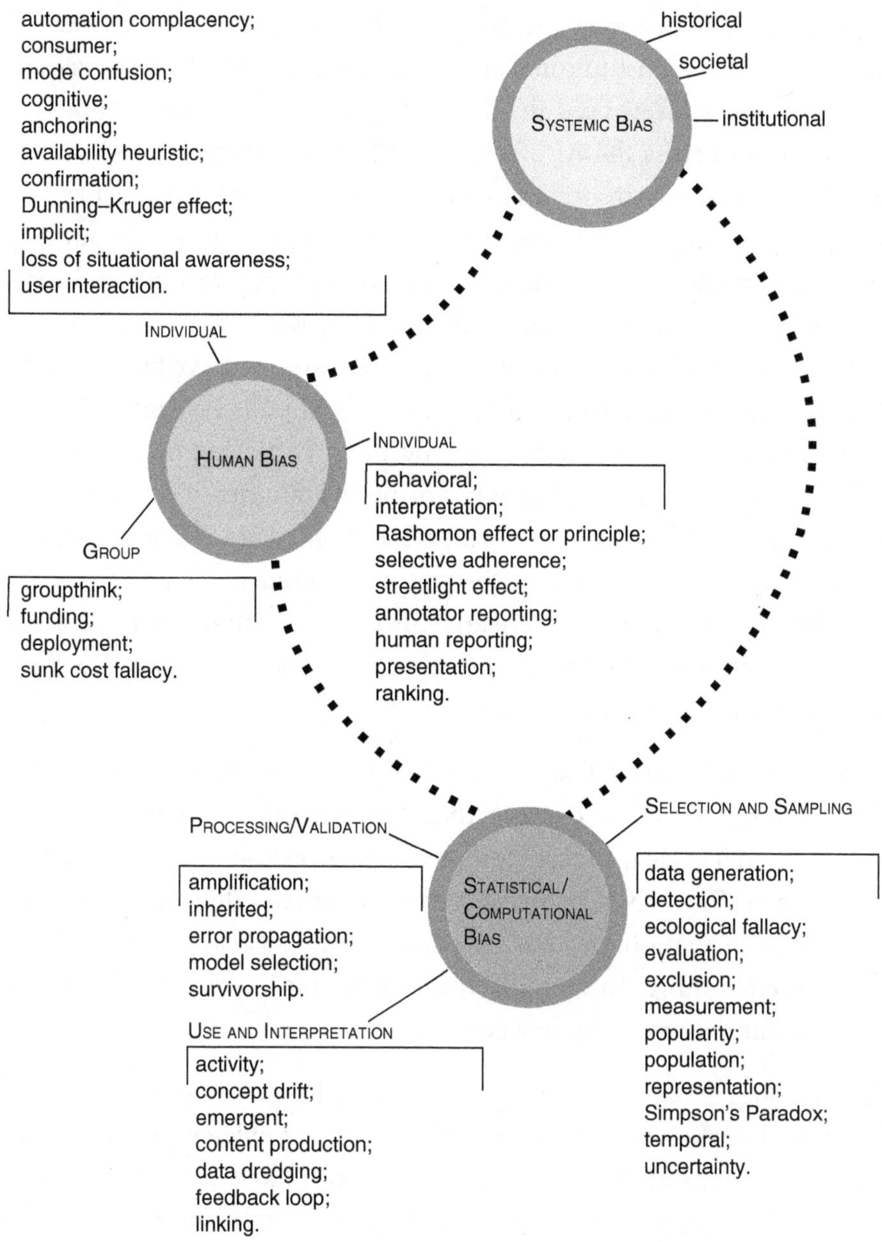

Figure 1: Categories of AI Bias

NIST's March 2022 paper on managing bias in AI notes, "Bias is neither new nor unique to AI and it is not possible to achieve zero risk of bias in an

---

Publication 1270, Mar. 2022, https://www.nist.gov/publications/towards-standard-identifying-and-managing-bias-artificial-intelligence, Figure 2.

AI system."[22] The first major AI bias that NIST describes it called systemic bias, also known as institutional or historical bias finding, "These biases are present in the datasets used in AI, and the institutional norms, practices, and processes across the AI lifecycle and in broader culture and society." Racial or gender bias underlies historical decisions and can affect the way AI makes predictions or decisions regarding people now. For example, an AI used to make decisions in human resources may recommend that a woman be paid less than a man with identical work skills, but that decision may be based on the historical fact that women have been paid less than men for the same work. In this case, the AI pulls systemic, historical discrimination into its decision-making process, biasing the recommendation. AI tends to be trained on enormous amounts of historical data, but we know that history comes with biases that need to be accounted for.

The human mind exists with its own inherent bugs and biases. As the NIST bias paper states, there are entire fields of study founded on the heuristics and biases in human thought and decision-making. The figure above notes several of these, including,

- the sunk cost fallacy, where people are reluctant to abandon a course of action because they have invested heavily in it, even when it is clear that stopping further investment would be better;
- anchoring, where people place unwarranted value on the first piece of information they receive when making decisions; and
- the Dunning–Kruger effect, where people overestimate their own abilities to perform in a certain area.

The Dunning–Kruger effect can be a problem for judges who may believe they can almost always predict the chance of a convicted person's return to prison, when statistics demonstrate that they likely cannot. Social science has documented dozens of standard glitches of the human brain that bring bias into decision-making, and these biases can be built into AI decisioning systems as well. AI systems do not think. However, they are developed, trained, and tested by humans, who are likely to impart some of their thinking biases into the predictive algorithms.

And finally, the above figure includes statistical and computational biases that can occur when the AI training data is not representative of

22. *Id.* at p. ii.

the population the AI is expected to make predictions about. The NIST authors explain, "These biases . . . can occur in the absence of prejudice, partiality, or discriminatory intent. In AI systems, these biases are present in the datasets and algorithmic processes used in the development of AI applications, and often arise when algorithms are trained on one type of data and cannot extrapolate beyond those data."[23] There are multiple ways in which these computational biases can arise, but any of them, or any combination of them, can skew the predictive output of a decisioning AI.

All of these systemic, human, and computation problems can create bias in a predictive AI, and the AI's developers must be aware of all of them and must test for the effects of potential bias. This can be clearly seen when applied to sentencing and parole decisions. Where an AI is assisting in decisions affecting years of a person's life, we would like to know that the AI was developed correctly and minimized or eliminated the many biases that could lead to faulty outputs. If the AI takes all of the relevant information into account[24] and assigns a number in evaluation of a convicted person's life options, we expect that number to be calculated fairly and not be the result of bias, whether or not that bias would be considered unlawful. We make the assumption that a machine learning calculation

23. *Id.* at p. 9, citing, Organization of Scientific Area Committees for Forensic Science, "OSAC Preferred Terms," 2021.

24. Input selection provides a similarly significant problem for creating accurate predictions about future behavior of humans. How does the human designer of the prediction machine choose which inputs have value and which should be ignored? Which inputs deserve more weight than others in the predictive calculations? As described throughout this chapter, humans are complex beings with variable personalities, changing priorities, and free will; choosing which factors to consider about personal history or current preferences can be the most difficult part of the predictive task. Which parts of a person's life or personality are predictive of future actions? Does a prior arrest show a propensity toward behavior on the edge of the law, or will it be the catalyst that pushes a person to avoid risky activity? Does hostility to authority show a desire to strike against the system or a longing to be left alone—or both?

Input selection is a different, and equally important, issue to the limits of AI training and the restricted data that the AI model uses to produce output. I am not including this discussion in the main body of the text because this set of input questions tends to be an entirely human endeavor, as opposed to the workings of the decisioning AI model once the baseline information is entered. If we are assuming that a decisioning AI model's creator has selected all of the right (and none of the wrong) factors to anchor its predictions, then we may still be wrong about the effectiveness and scientific basis of the AI model's output.

is scientifically sound, when, as NIST shows us, there are many ways the predictions and decisions may go awry.

Aside from bias, another problem in using AI models to assist people-centric decisions is the narrow focus of the AI models, which only have the information provided to them and no more. I have discussed in depth the problems with AI's lack of context. The AI can provide a certain analytical bent to the information at its disposal, but it does not have the context and common sense that even a human child might possess. This can lead to poor decision suggestions. *Wired* magazine provides a thoughtful example involving the driving recommendations supplied by an AI-base mapping program.

> Heading from Springfield, Missouri, to Clarksville, Arkansas, navigation apps recommend the Arkansas 43. While this can be the fastest route, the GPS's algorithm does not concern itself with factors important to truckers carrying a heavy load, such as the 43's 1,300-foot elevation drop over four miles with two sharp turns. The road once hosted few 18-wheelers, but the last two and half years have seen a noticeable increase in truck traffic—and wrecks. Locals who have watched accidents increase think it is only a matter of time before someone is seriously hurt, or worse.
>
> Truckers familiar with the region know that Highway 7 is a safer route. However, the algorithm creating the route recommendation does not. Lacking broader insight, the GPS only considers factors programmed to be important. Ultimately, the algorithm paints an incomplete or distorted picture that can cause unsuspecting drivers to lose control of their vehicles.[25]

In other words, an AI decision-making algorithm is only as good as the information that goes into it. One potential problem is the quality of data the model is trained on. Another is the quality of analysis that humans exhibit when choosing which data will be important for the ultimate decision. In choosing sentencing or parole options, maybe childhood exposure to violence would be an important factor in predicting future violent crime commission. Do any (or all) of the relevant AI models take this into

---

25. Jason Tashea, *Courts Are Using AI to Sentence Criminals. That Must Stop Now, Opinion: Courts should pause the use of algorithms for criminal sentencing*, Wired, Apr. 17, 2017.

account? And if so, do we understand directionality? Would childhood violence exposure correlate with more violence committed in later life, or less? Would the correlation stop after age 30? After age 60? And how would this factor relate to homelessness, previous convictions for violent crime, time in prison, associations with known violent felons, or adult domestic violence victimization? Would the designers of a predictive AI to assist in criminal justice sentencing know any or all of these answers? It seems like the people creating these systems would need to constantly update their knowledge of academic social science research studying criminal recidivism so that the predictive AI model could adjust for more important facts.

Given that algorithmic decision-making has been happening for decades, and that we understand many of the flaws that can make such decisions ineffective, it was only a matter of time before policymakers tried to protect the general public from the effects of these decisions. In addition, never underestimate the human concern about "creepiness" and dehumanization from granting machines influence over important aspects of our lives. Dehumanization has been an important social concern since the dawn of industrialization and can be significant in the algorithmic decision-making discussion "because it does not depend on the outcome of debates over whether hiring algorithms are biased. Future technical and ethical advances may address problems of bias and discrimination (citation omitted), but worries about dehumanization seem to be largely independent of these other issues."[26] Splitting from Hannah Arendt's work on dehumanizing social mechanisms, the decisioning AI debate is less about treating classes of people as "subhuman," and more about removing humans from an important aspect of life that once was completely dominated by humans, thus exorcising human judgment, human bonding, and human empathy from important decisions.

As stated in the context of algorithmic hiring decisions, "In human relationships, our values and the relations between our values are often complex and difficult to balance. Any attempt to impose artificial value clarity on this cascade of values by means of predictive algorithms—even ones that have been subjected to rigorous validation methods—is bound

---

26. Megan Fritts & Frank Cabrera, *AI Recruitment Algorithms and the Dehumanization Problem*, 23(4) Ethics & Info. Tech. 791–801 (2021).

to strike some [people] as uncomfortably alienating. . . ."[27] When the U.S. government began tracking social security numbers in 1936, people fretted about the dehumanizing step of replacing people's names with a government-issued number. There are excellent administrative reasons for doing so, but numbering citizens seems like a big step down a slippery slope where human rights are less important than accounting. Similarly, allowing important life decisions like criminal sentencing, job prospecting, and loan offers to be managed, or even heavily influenced, by machines rather than humans seems like a leap toward a future where people matter less to society and less to each other.

For all of these reasons, politicians have been promising to keep "a human in the loop" for major decisions as companies and governments move further down the road of implementing algorithmic decision-making. Some of the first broad laws to curb the effect of AI on people in Western societies have restricted the unfettered use of decisioning AI models. For example, in 2012, President Barack Obama proposed the Consumer Privacy Bill of Rights so people could correct information used by algorithms in a similar way to changing incorrect information in a credit report. The proposal never received congressional approval. The U.S. Congress has been increasing attention on AI regulations, as noted in a 2024 Stanford survey, there has been "a sharp increase in the total number of proposed bills that relate to AI from 2015 to 2021," and that the 117th Congress "is on track to record the greatest number of AI-related mentions since 2001, with 295 mentions by the end of 2021, halfway through the session. . . ."[28] Similarly, the same study shows the number of U.S. state-level bills on AI increasing from almost none in 2016 to 150 state AI-mentioning bills proposed and 38 passed in 2023, with California and Maryland leading the way.[29] So addressing the perceived societal problems with AI has clearly been a priority for U.S. lawmakers.

But the United States was not to be the initial leader to protect its citizens from algorithms. The European Union (EU) passed an influential limitation on AI as part of the General Data Protection Regulation (GDPR) effective May 28, 2018. The GDPR is primarily an update to EU

27. *Id.*

28. *The AI Index Annual Report: Measuring trends in AI*, Stanford University Human-Centered Artificial Intelligence, Apr. 2024, Chap. 7.

29. *Id.* at p. 18.

privacy laws that had been in force since 1995, providing more specific regulations and more biting sanctions for violations, but it also included a right for an individual to not be subject to a decision which produces legal effects based solely on automated processing.[30] This right could be exercised against businesses, governments, or other entities using AI decision-making to affect the lives of people residing in Europe. The GDPR was defining the rights of a class of people it called "data subjects" and they could be subject to AI making decisions based on their data just as they could be harmed by the misuse of their sensitive personal data. Article 22 provides, "The data subject shall have the right not to be subject to a decision based solely on automated processing, including profiling, which produces legal effects concerning him or her. . . ."[31] While there are limited exceptions to this rule written into the Article itself, even for application of most of those exceptions the data subject must be able to request that a human participate meaningfully in the decision at issue. An exclusively AI-based decision is unfavored.

GDPR Article 22 has already generated impactful decisions. For example, on July 12, 2023, the Court of Justice of the European Union held that creating a credit score constitutes an "automated individual decision" within the meaning of Article 22, and this decision holds the potential to upend the lending and mortgage market in Europe. This reflects a broad interpretation of the algorithmic decision-making rule. For decades, lenders have relied on credit scoring to assist in consumer loan decisions. If the credit scores are not available, then what calculations will replace them, and will lenders be less likely to extend credit to higher-risk applicants? However, the decision may only mean that lenders can continue to use credit scoring, but will need applicant consent to do so, and that credit applicants can ask for human consideration of the lending decision. On its face, the decision is only applicable in cases where the credit scoring system is "strongly" relied upon to make third-party financial decisions, which implies that credit scoring is not finished in Europe, but must be used alongside other considerations in relevant decisions that affect people's lives. The Hamburg Data Protection Authority cheered this decision, claiming that the calculated score "has consequences far beyond

30. Regulation (EU) 2016/679 (General Data Protection Regulation), Art. 22.

31. *Id.*, para. 1.

the scope of credit agencies – because it is transferable to the use of many AI systems."[32] Already, European data protection authorities have been considering how to apply the decision in cases where scoring systems may be used for organ recipients or criminal justice systems, seeking to provide European residents with more rights to question or reject the current methods of AI-enhanced decision-making.

The EU took this line of regulation even farther in 2023 with the passage of an entire set of legislation aimed at regulating the development and use of artificial intelligence. On 12 July 2024, the European Parliament published the European Union Artificial Intelligence Act[33] (EU AI Act), a comprehensive legal framework for the regulation of AI. For the purposes of this chapter, we must note that the primary concern of the European legislators seems to be AI models that make predictive decisions affecting EU residents. First, while the EU AI Act was passed in 2024, it was written years earlier, and due to the generally lengthy processes for drafting and accepting EU legislation, this Act does not specifically cover any aspect of generative AI, which have existed in simpler forms for most of a decade, but was not central to the AI discussions until the last months of 2022. Second, the EU AI Act divides AI models into risk categories, and classifies the following as "prohibited" AI systems, AI models that assess:

- "the risk of an individual committing criminal offenses solely based on profiling or personality traits, except when used to augment human assessments based on objective,"
- "verifiable facts directly linked to criminal activity," and
- AI models providing "social scoring, i.e., evaluating or classifying individuals or groups based on social behaviour or personal traits, causing detrimental or unfavourable treatment of those people."

Many variants of AI models could fall into one of these categories, including most lending programs used by financial institutions which classify individuals based on personal traits, often to the individual's detriment.

32. *Effects of the Schufa judgement on AI applications - automated decisions must not play a decisive role, Fuchs: "Judgement also of ground-breaking importance for AI-based decisions,"* Press Release from The Hamburg Commissioner for Data Protection and Freedom of Information, Jul. 12, 2023.

33. Regulation (EU) 2024/1689.

As we have seen, law enforcement has been using AI models to predict the risk of recidivism for many years, and now those programs seem to be prohibited in a major section of the Western world. Further, the EU AI Act also classifies some AI systems as "high-risk" models which require heightened governance and risk management to be allowed in production at all for the EU. The law states, "AI systems are always considered high-risk if it profiles individuals, i.e. automated processing of personal data to assess various aspects of a person's life, such as work performance, economic situation, health, preferences, interests, reliability, behaviour, location or movement." This covers many of the current use cases for decisioning AI which takes the history and responses of an individual (the profile) and uses this to predict the individual's eligibility for benefits or punishments.

While the United States has not passed a comprehensive nationwide AI legal framework, several jurisdictions in the United States have passed laws limiting the use of decisioning AI models in one context or another. Some of these laws address one aspect of highly specific circumstances in which AI decisioning models could be used. For example, the Illinois's Artificial Intelligence Video Interview Act[34] only covers "[a]n employer that asks applicants to record video interviews and uses an artificial intelligence analysis of the applicant-submitted videos" to evaluate job applicants. This Act requires that the interviewing employer provide notice and detailed explanation to the applicant of the AI tools to be used in the interview and to receive the applicant's consent for these tools to be used in the interview process. In addition to forced destruction of the videos, an employer using this AI-based technology for employment decisions "may not share applicant videos, except with persons whose expertise or technology is necessary in order to evaluate an applicant's fitness for a position." Finally, the Act also requires employers that solely rely upon an artificial intelligence analysis of a video interview to determine whether an applicant will be selected for an in-person interview to collect and report to Department of Commerce and Economic Opportunity of the state: "(1) the race and ethnicity of applicants who are and are not afforded the opportunity for an in-person interview after the use of artificial intelligence analysis; and (2) the race and ethnicity of applicants who are hired."

34. 820 ILCS 42, effective Jan. 1, 2020.

Clearly the Illinois legislature found AI video analysis of job interviews to be a risk for workers in their state, and crafted a highly particular legislative solution that both discouraged such AI decision-making and regulated its effects.

Other state and local legislatures were also checking the use of AI-enabled technology in job interviews.[35] For example, New York City passed a measure, effective in 2023, restricting employers from using "automated employment decision tool to screen a candidate or employee for an employment decision."[36] New York City requires that employers conduct annual bias audits on their automated employment decision tools, and to make the results of these bias audits publicly available on the employer's website. At least ten days before the tool is used on an applicant or employee, the applicant or employee must (1) be given notice and description of the tool, (2) be given a description of the job qualifications that the tool will be used to assess, and (3) must have an opportunity to request that a different evaluative path be taken. The breadth of this regulation is interesting because the most popular automated employment decision tools are those that winnow down a set of 1000 qualifying resumes to a manageable number, like the top 10 or 20, so that a human resources manager doesn't need to spend entire days reviewing questionable resumes. It seems that under plain text of the New York City regulation, each of those 1000 applicants—whether or not they knew they were being considered for a specific job—would need to be notified ten days ahead of time about use of the tool, the job description, and the type of information collected for the tool, even for a "first blush" quick evaluation of piles of resumes. Further, a \$500–\$1500 penalty may attach to each of the 1000 applicants not given the ten-day notice. This might have been the City's intent for limiting use of these tools, but more likely the drafters cast a wider net than practical in attempting to protect New York City job applicants from being evaluated by decisioning AI without their permission.

The recent and broadest law to limit and regulate AI decision-making is the Colorado Artificial Intelligence Act (Colorado AI Act), passed in

35. Around the same time (effective Oct. 1, 2020), Maryland enacted a law prohibiting employers for using facial recognition technologies during interviews without receiving consent of the person being interviewed. Md. Code Ann., Lab. & Empl. § 3-717.

36. N.Y.C. Admin. Code §§ 20-870–20-874.

2024.[37] Colorado takes the European strategy of a comprehensive risk-based consumer protection legal framework affecting those who develop AI models and those who engage those models in ways that affect Colorado residents. This law was passed to remedy a concern over the possibility of "algorithmic discrimination" which is defined as "an unlawful differential treatment or impact that disfavors an individual or group of individuals on the basis of their actual or perceived age, color, disability, ethnicity, genetic information, limited proficiency in the English language, national origin, race, religion, reproductive health, sex, veteran status, or other classification protected under the laws of this state or federal law."[38] The law is set to go into effect in February 2026, although within days of passage, Colorado's governor and key legislators were already discussing amending the newly-signed AI restrictions.

This law charges developers and deployers to protect consumers (using reasonable care)[39] from any known or reasonably foreseeable risks of "algorithmic discrimination" arising from the intended uses of decisioning AI in certain fields that the law defines as "high-risk AI systems." This is a system that makes, or is a substantial factor in making, a "consequential decision," which is a decision that has a material legal effect on the provision or denial to any consumer of a number of already-regulated decisions that businesses and other entities make about consumers, including housing, lending, insurance, education, and health care.[40] In one sense, key aspects of the Colorado law only regulate decisions that are already against the law, but now these decisions are especially disfavored under Colorado law because they were made by decisioning AI. Federal and state laws proscribe both disparate treatment and disparate impact discrimination in housing, lending, insurance, education, and health care, among other areas. The primary evil being banished by this law is "algorithmic discrimination," which, as defined in the Colorado AI Act, only includes "unlawful differential treatment or impact that disfavors an

37. Colo. Rev. Stat. § 6-1-1701.

38. *Id.* at § 6-1-1701(1)(a).

39. The duty of care includes a number of consumer notifications, ongoing programs, and system analysis reports.

40. Colo. Rev. Stat. § 6-1-1701(3) & 6-1-1701(9). The full list is education enrollment or opportunity, employment or an employment opportunity, a financial or lending service, an essential government service, healthcare services, housing, insurance, and legal services.

individual or group of individuals" on the basis of protected traits.[41] So the actions only qualify for censure under this law if they are already unlawful. There is a great deal of regulation in this law that will be expensive and time-consuming for any companies implementing decisioning AI for their businesses in Colorado, and the benefits to Colorado consumers seems small given that the prohibited activity was already prohibited by other laws.

Colorado AI model deployers must develop and apply a new risk management program for AI decision-making, and provide consumers with notice that an AI decision-making system is being used. The latter obligation includes statements about the purpose of the system, the nature of its decisions, and a notice of consumers' right to opt out of processing. In cases where the consumer is on the wrong side of a decision facilitated by AI, the system deployer must include a clear, effective statement disclosing the reasons for the decision and the data used to make the decision. The system deployer must provide both an opportunity to appeal the decision and to correct any erroneous data. Annual data impact assessments for each decision-making AI system are required for anyone deploying them in Colorado. Entities using AI decisioning systems must also make a publicly available statement summarizing the types of high-risk systems that the deployer currently deploys, how the deployer manages any known or reasonably foreseeable risks of algorithmic discrimination that may arise from deployment of each of these high-risk systems, and the nature, source, and extent of the information collected and used by the deployer.

In addition, the deployer must report on itself to the state attorney general and admit publicly that its decisioning AI unlawfully discriminated. This admission must be made within ninety days of discovering the discrimination. One could reasonably question why companies or other entities would want to begin implementing AI decisioning systems in Colorado, given that they would not need to jump through all of these administrative, operational, and regulatory hoops to apply their tools in other states. It is possible that Colorado is either attempting to discourage

41. Colo. Rev. Stat. § 6-1-1701(1)(a). Those protected categories are their actual or perceived age, color, disability, ethnicity, genetic information, limited proficiency in the English language, national origin, race, religion, reproductive health, sex, veteran status, or other classification protected under the laws of this state or federal law.

the use of AI in any kind of decisions involving consumers, or that it is trying to steal the "first mover" mantle of California, starting a trend of comprehensive state AI consumer protection regulations. In any case, if the Colorado AI Act is implemented in 2026 in its current form, it may shape the way AI is implemented in consumer contexts in the United States.

Colorado is also an early leader in incorporating into their omnibus personal privacy laws, a consumer right to opt out of AI decision-making. The 2021 Colorado Privacy Act includes an opt-out right from profiling resulting in "legal or similarly significant effects concerning a consumer." Profiling is very broadly defined as "any form of automated processing of personal data to evaluate, analyze, or predict personal aspects concerning an identified or identifiable individual's economic situation, health, personal preferences, interests, reliability, behavior, location, or movements."[42] Once again, to provide a practical option to refuse processing, a business would need to notify the consumer that it was planning to use such automated personal data processing.

Illinois passed the Artificial Intelligence Employment Act hot on the heels of Colorado AI Act.[43] This law amends the 1979 Illinois Human Rights Act, and creates a civil rights violation for an employer to use artificial intelligence that results in unlawful discrimination, and a separate civil rights violation to fail to provide notice to an employee that the employer is using artificial intelligence for employment decisions. Like Colorado, Illinois found it significant to add a new set of violations for using AI to do something that was already illegal. This is not an unheard-of legal philosophy: for example, committing murder is a crime, while committing murder with a gun can be either an aggravated version of the crime or an entirely new crime altogether. So violating civil rights with illegal employment decisions is a crime in Illinois, and now doing so through the use of AI is an extra civil rights violation. The Illinois Act broadly defines "Artificial Intelligence" as both generative artificial intelligence and as "a machine-based system that, for explicit or implicit objectives, infers, from the input it receives, how to generate outputs such as predictions, recommendations, or decisions that can influence physical or virtual environments." So nearly any type of AI functionality would qualify if the AI

42. Colo. Rev. Stat. §§ 6-1-1303(20) & 6-1-1306(6).

43. Illinois Public Bill 103-0804.

participated in interactions or decisions relating to employees who are Illinois residents. The Colorado AI Act is broader than the Illinois Act, with Colorado applying its rules to protect consumers of all types and Illinois protecting the civil rights of employees. Also the Illinois Act does not require affirmative action from an employer to assess, report, or mitigate the risks arising from the use of AI systems, as the Colorado Act demands.

The omnibus California Consumer Privacy Act (CCPA) also provides an option for consumers to remove themselves from automated decision-making practices.[44] California has written draft regulations for its right to be exempted from automated decision-making, but as of this writing it has not implemented them. The regulations are detailed and require businesses that use automated decision-making to provide notices to affected consumers. Those notices must "inform consumers about the business's use of automated decision-making technology and consumers' rights to opt-out of, and to access information about, the business's use of automated decision-making technology."[45] Several other states have these consumer options for automated decisioning built into their versions of comprehensive consumer personal privacy protection acts, and I expect many more will follow as the trend continues for new states to embrace European-style data protection laws.[46] It is likely that before the federal government makes this option available to U.S. consumers, all or nearly all of those consumers will already have such rights granted under state law. While there is no guarantee that a second decision—more human-driven and less automated—would turn out any better for a consumer who opts out

44. Cal. Civ. Code § 1798.185(a)(16), although since the definition of "consumer" under the CCPA also includes employees, California extends the right to opt out from automated decisions in hiring, termination, and other employment realms as well.

45. Proposed Automated Decisionmaking Technology Regulations, § 7017(a).

46. For example, see the Montana Consumer Data Protection Act (Mont. Code Ann. § 30-14-28), which enables individuals to opt out of "profiling in furtherance of automated decisions that produce legal or similarly significant effects" concerning the consumer; the Indiana Consumer Data Protection Act (Ind. Code § 1-24-15), which provides for individuals to opt out of "profiling in furtherance of decisions that produce legal or similarly significant effects;" and the Connecticut Data Privacy Act (Conn. Gen. Stat. § 42-518), which provides for a consumer to "opt out of the processing of the personal data for purposes of [. . .] profiling in furtherance of *solely* automated decisions that produce legal or similarly significant effects concerning the consumer." (emphasis added) Connecticut limits the option to circumstances where the entity makes a consumer decision solely through automated means, unlike other states that limit even the material influence of automated predictions on otherwise human decisions.

of AI decision-making, the fact that the option exists provides incentive for consumers disappointed by official decisions under these state laws to accuse the company or entity providing disappointing news of excessive automation in their decision-making process. The company or entity would then need to defend its entire system of choices. It is unknown whether the additions of all of these administrative requirements will chill the further implementation of AI decisioning tools as these tools develop and improve, but implementing companies will receive significantly more scrutiny on these decisions than was contemplated even a couple of years earlier.

I started this chapter examining the decision-making AI being used in law enforcement and corrections and have ended it with laws that regulate AI decisions made about consumers. It is clear that the entire category of decisioning AI not only raises legally significant issues like bias and unlawful discrimination, but that legislatures in the United States and across the world have been concerned enough about the effects of AI decision-making that laws restricting it have been introduced, debated, and passed. While the buzz about AI in the early 2020s has been dominated by interest in generative AI, another form of AI—which may or may not have been developed using large language models—has been vital to several important fields of endeavor. Decisioning AI, and its algorithmic predecessors have been used for decades, and lawyers should think about this category as distinct and legally significant.

CHAPTER 5

Generative AI

England's Queen Elizabeth I was skeptical about the first stocking-knitting loom, and in 1589 she denied a royal patent to its inventor, William Lee. The Queen was apparently concerned for the job security of the many people in Tudor England making their living from hand knitting. Her instincts on this matter were well-founded, as the mechanized loom would soon replace the intricately clicking and twisting fingers of hand knitters as a primary economic force in clothing manufacture. "For two hundred years, the stocking frame reigned. With its counterpart for weaving, the handloom, it helped England become known around the world for finely produced textiles. . . . hand knitters were largely replaced by the stocking frame operator."[1] These hand-knitting artisans, an early creative class, were no match against a machine that could unceasingly generate similar products faster and cheaper.

Two hundred years later, knitting machines were more sophisticated, and working handlooms was an artisan's job. Automated factories were putting the artisans out of work, and the effects of Napoleonic wars were impoverishing more of the English lower classes, so some took to mob violence to protest their plight. The targets of violence were cruel mill owners, the mills themselves, and the mechanized looms that filled them. "A stocking frame, a machine for knitting socks that was famously smashed by Ned Ludd in 1779, provided a foundational moment and a name to

1. Brian Merchant, Blood in the Machine: The Origins of the Rebellion Against Big Tech 10 (Little, Brown and Company, Hachette Book Group) (2023).

the Luddites, disgruntled weavers and famous machine breakers who harassed capital in England between 1811 and 1817."[2] Producing clothing has been both an art form and an important economic force for humans as far back as we can measure. It should not be a surprise that introducing mechanization into this process not only threatened the livelihoods of an entire class of artisans, but could lead to violence in opposition to these disruptions. The Luddites smashed machines because it was a straightforward way of expressing rage at the way factory owners were changing their world and treating the workers. Eventually Luddites moved to murder and burning facilities. Thousands of soldiers were deployed to protect factories in the North of England and Parliament passed a law making machine-breaking a capital offense. Several Luddites were killed by soldiers and some were hanged for murder.

We use the term "Luddite" now to mean any person who aggressively resists implementation of the newest machinery, but it is important to remember that nearly every economically-important advance in technology displaces skilled workers and tends to tilt the balance of production further in favor of capital and away from skilled production. By now this dynamic is centuries old, and it keeps imposing itself on groups of workers who previously thought themselves invulnerable to technological changes. Thanks to the subject of this chapter—generative AI or creative AI[3]—an entirely new class of painters, photographers, videographers, musicians, writers and coders, and maybe other knowledge workers like lawyers, are seeing threats to their personal abilities to make a living from

2. Sofi Thanhauser, Worn: A People's History of Clothing 80 (Pantheon Books) (2022).

3. For purposes of this chapter, the term "creative AI" may be most accurate, because I am describing the AI functionality of generating an output in a form that had previously been the domain of writers, coders, musicians, and artists. As a term of art, "generative AI" performs these tasks and much more. Generative AI can produce all sorts of language-based interactions, sonic-based products, and visual imagery that is not threatening in any way to the creative classes. For example, the LLMs that serve as chatbots for some businesses to answer customer inquiries are correctly classified as generative AI, and they are outside the scope of this chapter. In this chapter, I intend to demonstrate how creative AI produces specific legal problems that are not inherent in the forms of AI discussed in other sole-category chapters, like decisioning AI and physical world AI. That said, generative AI is the standard accepted term for this type of tool, so to accede to the common understanding, I use that term in this chapter, although it is over-inclusive and less precise than creative AI. Despite my assertions in chapter 2 of the lawyer's need for precision, sometimes a default to the less-precise, more-common terminology is important to promote broader recognition of the concepts.

their own creativity and thought. Until 2022, the general public had not imagined that digital tools would come for the jobs of people who trade in words, sounds, and images.

Lawyers are not only potential casualties of the new word-generating technology, but the artistic class has already turned to their attorneys to find ways to minimize economic damage. Lawsuits have been filed by painters, songwriters, singers, actors, directors, and even newspapers against the intrusions of generative AI into creative businesses. The rules of such economic conflict will be decided in the courts, legislatures, and enforcement agencies. The U.S. Copyright and Patent Offices have already opined on the outputs of generative AI, and will continue to do so. The effects of generative AI have appeared in courtrooms, not just as subject matter of economic battles, but as court filings by counsel and likely as proposed evidence. The evidentiary possibilities are of special concern, as generative AI has the current capability to produce entirely fictitious audio, photo-like graphics, and video files that could be indistinguishable from similar files created by in-person recordings. This new technological reality may necessitate a change in court rules for accepting and authenticating evidence,[4] and could lead to admission of entirely faked evidence in some cases and effective discounting of entirely true evidence in others. It will most certainly cause significant evidentiary confusion in many cases that would not have happened without the wide availability of stunning faking capacity of generative AI. Believing your own eyes and ears will no longer be the comfortingly obvious option. This matters to lawyers not only as they argue court cases, but also as they evaluate situational risks for their clients.

What is generative AI? According to IBM, generative AI refers to deep-learning models that can generate high-quality text, images, and other content based on the data they were trained on.[5] These AI models had been trained by feeding millions of examples into the algorithm. For example, for a visual file generative AI like DALL-E from OpenAI, the model is trained on billions of visual images, and the model uses these examples to produce its outputs pixel by pixel. A user types a prompt into

4. Committee on Rules of Practice and Procedure of the Judicial Conference of the United States, Hon. John D. Bates, Chair, Report to the Standing Committee, Advisory Committee on Evidence Rules, May 15, 2024.

5. Kim Martineau, *What is generative AI*, IBM Research Explainer, https://research.ibm.com/blog/what-is-generative-AI, Aug. 17, 2024.

the request space, asking, for example, for a photorealistic image of an orange cat on a bookshelf. The model produces a series of four images that may be responsive to the prompt. The user can then choose an image to revise, or can refine the original prompt. At this writing, the primary public generative AI tools are provided in specific silos. Some generative AI tools generate text documents, some create sound files, and some provide visual files as output. It is likely that at the time of publication, these lines will be blurred and a user can create multiple file types with a single prompt to a single generative AI model.

In the late 2010s, generative AI captured the public's attention as a tool for modifying selfies and other social media pictures. In November 2022, ChatGPT-3 was released with fanfare and the world became engaged with the concept that a generative AI model could generate interesting and sometimes useful outputs with just a prompted phrase. DALL-E was soon producing French impressionist versions of teddy bear art and students were turning in AI-generated papers analyzing the religious symbolism in *Moby Dick*. Generative AI seems to excel at attacking scientific problems and in writing functioning computer code. Celebrity voices, from living or dead people, could be fed into generative AI and new text could be generated in those voices, with Franklin Roosevelt apparently selling Dunkin' doughnuts or John Wayne sounding like he was supporting a children's dance recital. The possibilities seemed limitless, which is both exciting and disturbing. Most disturbed were the artists, actors, singers, and writers whose work was being mimicked, and those who saw threats to their livelihoods and important aspects of the traditional training and learning process for young people learning creative professions.

The outputs of generative AI can be shockingly similar to the forms the models are mimicking. This seems especially true when we first see and hear these outputs and have no comparatives beyond similar items generated by human brains or captured from the real world.[6] But we

6. Eventually, when people have many generative AI output examples to compare to human created or reality-captured examples, the generative AI results can be isolated and identified. At this writing, AI-generated "photographs" and other pictures show particular problems with producing realistic hands and feet, and creative story writing tends to rely on generic phrasing. *See*, for example, Curtis Sittenfeld and Susannah Meadows, *Can You Tell Which Short Story ChatGPT Wrote?*, N.Y. Times, Aug. 28, 2024, where Sittenfeld says of the AI-generated story, "I feel like if you

can easily misunderstand what is being produced. It has been said that earlier forms of AI excel at pattern recognition, while generative AI excels at pattern creation.[7] Generative AI takes the patterns it has been trained to recognize—a French impressionist landscape, a specific person's vocal patterns, and the form of a motion for summary judgment in court—and it generates output that predicts how the prompt would appear in that form. What would it look like if Claude Monet painted the current Abu Dhabi skyline? What would it sound like if Bob Dylan sang *Old Town Road*? Needless to say, there is neither human creativity nor factual veracity built into the processing of these outputs. The generative AI production may be what it would sound like if Bob Dylan sang *Old Town Road*, but he never recorded this, probably has never sung it, and the output is nothing more than a pattern simulation. Lawyers who do not understand this are doomed to misuse the technology and fail their clients.

This is why the fined *Avianca* lawyers were fooled into thinking that their generative AI was producing a real legal document that could be filed in court: they did not understand what the machine was doing to produce its output. The generative AI was not a standard search engine that scoured the web for the answers to your question and then built those answers into a usable document. Instead, the generative AI simply predicts what a prompted document might look like. And in the case of the famously flawed *Avianca* filing, the generative AI had trained on court filings. Court filings tend to include legal citations to support the arguments. The generative AI mimicked the citations so that they mostly appeared to be real citations to real cases, but they were just generated nonsense that looked the part—a Potemkin village of a legal memorandum with an impressive surface but it does not hold up to any serious scrutiny.

---

plucked an individual sentence from ChatGPT's story, it's probably fine, or even better. It's probably smoothly written, but cumulatively, something is missing. There's not any real emotion or anything that makes it feel very specific."

"Because stories generated by AI models can only draw from the data that those models have been trained on, those produced in the study were less distinctive than the ideas the human participants came up with entirely on their own." Rhiannon Williams, *AI can make you more creative—but it has limits; Although it can boost individuals' creativity, it seems to homogenize and flatten our collective output*, MIT TECH. REV., Jul. 12, 2024.

7. Bernard Marr, *The Difference Between Generative AI And Traditional AI: An Easy Explanation For Anyone*, FORBES ONLINE, Jul. 24, 2023.

As generative AI becomes a more important tool in our society, understanding its functionality and limitations will be important for nearly every kind of lawyer. Litigators can use the word-generating power of the technology, but will always need to check, edit, and reword all documents so that every aspect is supported and argued correctly. But litigators will also be faced with generative AI deepfakes as evidence, or the question of whether evidence submitted against their clients is real, and for them, understanding the ways that evidence can be generated may prove the key to protecting a client from disaster. Transactional lawyers are equally likely to use generative AI to produce documents and have the same obligations for veracity and cogency. Intellectual property lawyers are presented with AI-generated inventions and writings with complicated protection regimes which involve what aspects of the work or invention were created by humans and what aspects were generated by the AI. But many current legal cases question the method of creating generative AI models, rather than focusing on their outputs.

Current generative AI models were trained by using the massive amounts of information available on the internet. From a lawyer's perspective, the commercial internet is based on an underlying set of permissions. Companies, governments, people, and entities create websites and post them on servers that can be accessed over networks by everyone with a browser. Those browsing people operate under an implied right to view or listen to anything available on the open web. If you don't want us to see it, don't post it, or at least put it behind a firewall. Otherwise, the general public holds a right to view it, and probably a right to copy it for their own purposes. Maybe that just means copying a link, but it can also mean making a screenshot or copying the site for non-commercial reasons. After 30 years of internet-based legal precedent, we now know that we do not necessarily have implied permission to copy all or significant part of a website and to post it back on the internet for our commercial purposes.[8] But we should have rights to view a site and make some sort of transformative use of the content that doesn't directly harm the original website controller.

---

8. Websites are considered "works of authorship fixed in any tangible medium of expression" and are thus eligible for copyright protection under 17 U.S.C. § 102(a). See *Ticketmaster LLC* v. *RMG Technologies, Inc.*, 507 F. Supp. 2d 1096, 1104 (C.D. Cal. 2007).

The present crop of generative AI is an outgrowth of these internet permissions. To function in a useful manner, large language models (LLMs) need to train in oceans of language. Audio and visual machine learning systems operate with sophisticated learning algorithms also trained on enormous streams of data. If you want your generative AI to learn about images, then train it with images on the internet that are adequately numerous for contrasts, comparisons, and groupings—, they seem to be free. How many pictures of cats does an AI model need to absorb before it can effectively identify a cat from a dog (or a spotted housecat from a leopard or jaguar) and generate a requested housecat image? Hundreds of millions, and maybe more. Where else, except the internet, can the makers of generative AI find as much written language, as many images, and as much audio or video? The other realistic choices, if they even have enough samples to be useful, would cost significant money to acquire the right licenses. So nearly every major generative AI product now in public use was trained on at least some information gathered from the internet, and most of them were trained almost exclusively on such content.

This may be a problem. While caselaw demonstrates that browsers clearly have an implied right to view content publicly available on the web,[9] they are less clear about what other rights we can exercise with this content. So sending an AI-developing algorithm through public websites for training is currently an uncertain right. To analyze whether internet scraping to train generative AI models violates copyright law, first we should know about how this training takes place. When the AI development tool stops at a website of interest, does it somehow copy or ingest the content from the site, or is it able to train by simply viewing the content? The act of copying is more likely to trigger liability under the copyright laws. Second, we should ask what the generative AI model is being trained to do. If the generative AI model is simply trained to answer general questions in English, then the fact that the model has been trained on half a million words of a specific author's books should not matter. However, if the generative AI is being trained to write new books that mimic the language choice, character development, and themes of the same author, its work might be actionable. If the generative AI was trained to write new

9. *Field* v. *Google Inc.*, 412 F. Supp. 2d 1106 (D. Nev. 2006).

books containing the main characters of the author's older novels, then such a tool is almost certainly committing copyright violations.[10]

After an initial misfire, a group of artists sued generative AI start-ups like Midjourney and Stability AI alleging that these AI companies violated laws by training AI image models on five billion images including the plaintiff artists' work without consent.[11] The case may be decided by the time this book is published, but the legal arguments underlying the case demonstrate why generative AI presents unique legal problems not raised by other AI functionality. The artists have claimed that they are damaged because the generative AI models trained on the artists' work produced output that unlawfully resembles the work of these artists. In other words, a person using the generative AI tools of the defendants can create pictures in the distinctive style of the plaintiff artists, without even needing to ask or pay the artist to create work in their signature style. As stated in *The Washington Post* article about this case, "At stake is the business model behind a generative AI boom built largely on the free, unauthorized use of works of all kinds from across the internet to train sophisticated software capable of producing humanlike works of its own. . . . A key question, as lawsuits move forward, will be whether the use of such material to train AI models constitutes 'fair use' or illegal infringement."[12] I would expect that the answer, after much litigation on many creative fronts, will be that training an AI on material freely available on the internet is not "per se" a copyright violation, but it could be a violation depending on how the material is turned into output by the particular AI model.

---

10. While entire volumes of text have been developed as fan fiction, shipping certain characters from widely published authors or other fictional works, and creating new story lines, unfortunately much of this work likely infringes on the copyright of the original characters' creators, should those creators choose to aggressively protect their work. It is widely recognized in the United States that fictional characters can gain copyright protect apart from the work they appear in. *See*, Paul Goldstein, Copyright: Principles, Law and Practice 158 (OUP) (1989). For example, the characters of Rocky Balboa and James Bond (see, *Anderson* v. *Stallone*, 11 U.S.P.Q.2D 1161 (C.D. Cal. 1989) and *Metro-Goldwyn-Mayer* v. *American Honda Motor Co.*, 900 F. Supp. 1287 (C.D. Cal. 1995)) have been held to be protectable under copyright law, while other well-known fictional characters like Sam Spade, have not (*Warner Bros. Pictures* v. *Columbia Broadcasting*, 216 F.2d 945, 950 (9th Cir. 1954)).

11. *Andersen* v. *Stability AI Ltd.*, No. 23-cv-00201-WHO (N.D. Cal. Aug. 12, 2024).

12. Will Oremus, *AI's legal reckoning is one step closer*, The Washington Post, Aug. 14, 2024.

In the artists' case against Stability AI, the court seem to lean toward this position. The judge allowed the most important copyright claims to advance, and he cast aside the defense arguments that would have shielded them from liability no matter how closely the AI outputs matched the copyrighted work of an artist. But he also discussed how the plaintiffs would need to show more than the bare minimum claim of "this AI was trained on my work" in order to succeed in their case. "[H]is opinion is also realistic about what the plaintiffs will ultimately have to show. It won't be enough for them to say the AI was trained on their works; they'll probably need to show that the AI is actually imitating their specific works."[13] In other words, training AI on freely-available internet content is likely to be classified as fair use of the copyrighted work unless the AI uses that content to produce something that harms or otherwise reasonably offends the copyright holder. Harm might be suffered if the generative AI model produces work too closely resembling the work of the copyright holder, and would certainly be suffered if the AI model simply reproduces work, pixel-by-pixel, that it has been trained on. Visual artists can likely show real damage to their careers if potential patrons could reproduce the artist's work with a well-worded prompt into a generative AI model.

The cases in this field receiving understandably more media attention than the visual artists are those cases filed by media companies themselves against major generative AI companies. These media lawsuits claim that training a generative AI model on a newspaper's copyrighted (and firewall protected) work product violates the plaintiffs' copyrights, and that either summarizing the news and feature articles or reprinting significant parts of them entirely is another set of copyright violation. Among other things, the plaintiff media companies claim that tech firms need permission to perform these acts. Other media companies who are not involved in these lawsuits have charged generative AI companies substantial sums for the rights to train on a library of published data. Given that OpenAI's chairman Sam Altman has said in official statements, "it would be impossible to train today's leading AI models without using copyrighted materials,"[14]

---

13. *Id.*, quoting Cornell University law professor James Grimmelmann.

14. OpenAI submission to the UK House of Lords Communications and digital select committee, described in Dan Milmo, *'Impossible' to create AI tools like ChatGPT without copyrighted material, OpenAI says: Pressure grows on artificial intelligence firms over the content used to train their products*, The Guardian, Jan. 8, 2024.

payments for access to their libraries may drive up costs for the next generation of AI models.

The New York Times lawsuit accused Microsoft and OpenAI of taking a "free-ride on The Times's massive investment in its journalism" and that "There is nothing 'transformative' about using The Times's content without payment to create products that substitute for The Times and steal audiences away from it."[15] As observed above, the most compelling arguments in this complaint concern how the copyrighted information is intended to be used by the generative AI companies, rather than those based on the simple act of training the generative AI model on copyrighted material. The stakes in this case are unfathomably high for the generative AI companies, as *The New York Times* has not only asked for an undetermined amount of money in the billions of dollars, but also that generative AI models trained on its copyrighted material be destroyed. Destruction of an illegally-trained AI has already been recognized as a valid outcome in similar cases,[16] so OpenAI and its fellow defendants are playing an existential game of chicken with *The New York Times.* If *The New York Times* wins and a court awards all its requested relief in the matter, the AI companies may need to scrap the products that they have spent billions to create.

Some of these lawsuits include contract claims. Website owners often rely on posted terms of use to try to place some limit on the type of activity that browsers can conduct with the content on their sites. Courts tend to uphold the behavior limitations of online terms of use,[17] though not in all cases.[18] At the moment, but maybe not for long, generative AI companies have been training their LLMs on the content of websites, before nearly any website users knew, to build specifically-targeted generative AI prohibitions onto their terms of use. Including proscriptions from gathering

15. *The N.Y. Times Co.* v. *Microsoft Corp.*, No. 23-cv-11195 (SHS) (S.D.N.Y. Aug. 6, 2024).

16. There is some precedent for this drastic penalty. In 2021, the U.S. Federal Trade Commission reached settlements with Everalbum and WW International, Inc. (formerly Weight Watchers) that included requirements that AI trained on improperly obtained and/or used data must be destroyed. *In the Matter of Everalbum, Inc.*, Federal Trade Commission, Decision and Order, Docket Number C-4743, May 6, 2021; *US* v. *Kurbo, Inc.*, United States District Court Northern District of California, Case No.: 22-CV-946, Joint Motion for Entry of Proposed Stipulated Order, Feb. 16, 2022.

17. *Meyer* v. *Uber Technologies, Inc.*, 868 F.3d 66 (2d Cir. 2017).

18. *Long* v. *Provide Commerce, Inc.*, 245 Cal.App.4th 855 (Cal. Ct. App. 2016).

a website's content to train a generative AI model could affect whether a court would find the practice of content capture for this purpose fair use under copyright law. There will undoubtedly be a series of cases analyzing the language from the contract terms of websites and whether that language adequately prohibited scanning of the site's content and using that content to train or test generative AI models.

The U.S. Copyright Act permits creators to use some parts of copyrighted work for certain specified reasons. This allows more flexibility for discourse around the copyrighted work of others. Section 107 of the Copyright Act provides the statutory framework for determining whether something is a fair use and identifies certain types of uses—such as criticism, comment, news reporting, teaching, scholarship, and research—as examples of activities that may qualify as fair use. This concept reconciles the Copyright Act with the First Amendment guarantee of free speech, and "it allows one to use and build upon prior works in a manner that does not unfairly deprive prior copyright owners of the right to control and benefit from their works."[19] Determining whether use of a copyrighted work is officially fair under the law is more of an art than science. A copyrighted work may be used for teaching—an acceptable activity—but still trigger copyright liability for using too much of the original material.[20] These are the factors for examining whether utilization of copyrighted material qualifies as fair use: the purpose and character of the use, the nature of the copyrighted work, the amount and substantiality of the portion taken, and the effect of the use upon the potential market for or value of the copyrighted work.

Clearly, these factors are loose and can present differently in each case. It is instructive to examine two recent internet copyright fair use cases that were decided in opposite results. In *Yang* v. *Mic Network Inc.*,[21] the defendant posted an article critical of a second article and included a screenshot of the second article. The court ruled that this posting was fair use because the defendant's article was a criticism of the first. On the other hand, in

19. Harvard University Office of the General Counsel, *Copyright and Fair Use*, "Fair Use of Copyrighted Material," https://ogc.harvard.edu/pages/copyright-and-fair-use, accessed Aug. 18, 2024.

20. *Cambridge University Press* v. *Becker*, 446 F. Supp. 3d 1145 (N.D. Ga. 2020).

21. Nos. 20-4097-cv, 20-4201-cv (2d Cir. Mar. 29, 2022).

Associated Press v. Meltwater U.S. Holdings, Inc.,[22] defendant Meltwater created a website crawler that pulled news articles from *Associated Press*, stored them in a database, categorized them, and allowed users to search the articles. The court found this was not fair use. *Associated Press* articles were factual in nature and were already published. These facts favored a finding of fair use. However, the other fair use factors were not in Meltwater's favor. Meltwater's activities seem in many ways similar to what *The New York Times* and others are complaining about when they suggest that the generative AI models from Microsoft and OpenAI have ingested copyrighted work from *The New York Times* and will be regurgitating that work in a way that takes economic value away from the newspaper. Just like the *Meltwater case*, the generative AI companies are providing third parties with copyrighted work without the need to subscribe to the newspaper it appeared in or see that newspaper's advertising.

This case is reminiscent of a similar copyright filing in the 1990s as the internet was becoming a browser-based medium. From the beginning of the commercial internet, the traditional newspaper industry worked to build business models that would allow them to continue to sponsor journalism. In the mid-1990s, many city newspapers had websites that competed with internet content from local television stations. But prior to the age of Google and easy-to-use search algorithms, it could be difficult to find a journalistic website if the human browser did not know the site's precise URL. To resolve this problem (and make money) a company called Total News created a website where you could be linked to the stories appearing in newspapers across the country. If you were traveling to Portland, Oregon, you could go onto Total News and read the local paper for restaurant and hotel reviews. If a building exploded in Houston, you could quickly find descriptions from the local papers on Total News, even if you lived far from Texas.

Total News essentially linked to each of these other newspaper sites but rather than simply sending your browser to the newspaper's servers, Total News would frame the desired local content with its own advertisements. According to the companies that sued to stop this business, Total News profited by selling advertising on top of the local journalism paid for by the linked newspapers. *The Washington Post* led a group of

22. 931 F. Supp. 2d 537 (S.D.N.Y. 2013).

litigants against this early internet business model and successfully shut the Total News site down.[23] A contemporaneous legal analysis found that "The complaint filed in Washington Post v. Total News describes a perfect example of how a low-cost entrant capitalized on the huge investments made by others to provide Internet content."[24] Total News claimed that it never copied the content of other sites, merely linking to them and laying its advertising frame over top. While this might have undercut the copyright case against Total News, *The Washington Post* and other plaintiffs alleged common law claims of stealing "hot news" and other value from them.[25]

The case was settled in June 1997, with Total News agreeing to remove its advertising-based framing from its links to the plaintiff's sites, now serving the function of a news-based aggregator with standard linking rights. One of the key claims driving this settlement was the charge that Total News was misappropriating advertising revenue that had been generated by "the sweat of the brow" from *The Washington Post* and other news sites. This claim is very different from copyrights claims and may be instructive to *The New York Times* or the artists suing OpenAI for training its LLMs on their work. "The policy goals of copyright law and misappropriation are different. Copyright law seeks to protect original expression while state misappropriation laws seek to protect labor."[26] So those companies seeking compensation for AI training might be best served focusing on how generative AI companies had been taking free rides on the extensive work and resources news organizations spent creating beautifully written stories about the interactions in the real world, rather than the fact that generative AI companies violated copyrights when they incorporated that

---

23. *Washington Post Co. v. Total News, Inc.*, No. 97 Civ. 1190 (PKL) (S.D.N.Y., filed Feb. 20, 1997).

24. Bruce P. Keller, *Condemned to Repeat the Past: The Reemergence of Misappropriation and Other Common Law Theories of Protection for Intellectual Property*, 11(2) HARV. J.L. & TECH. 401, 419 (Winter 1997).

25. *Id.* at p. 420, stating in footnote 96, "Not only does Total News itself acknowledge that some of the plaintiffs' sites qualify as 'breaking' news sites (it so describes them on its menu (footnote omitted)), it is now recognized that the immediacy of online publication permits it to 'scoop' traditional news outlets. *See* Elizabeth Weise, *Paper Scooped Itself on Web; Dallas Bomb Story Raises New Issues*, BERGEN RECORD, Mar. 2, 1997, at A18."

26. Frank C. Gomez, *Washington Post v. Total News, Inc.*, 13(1) BERKELEY TECH. L.J., Annual Review of Law and Technology 21, 33 (1998).

work for training an AI. The first case, based in competition law, seems more compelling than the second. Training AI on available text seems like it might be a reasonable fair use of the text, but taking another person's work and using that confiscation to compete against the working person seems unfair and should lead to compensation.

Would it make a difference on the "fair use" calculation if the website terms of use prohibited using the site's content to train generative AI models? In other words, in absence of a specific prohibition, generative AI companies may have the right to train their models on internet-available material, but once a website owner specifically prohibits such training the prohibition might remove the generative AI company's implied license to use the website in this manner. Effective prohibition language could have an effect on copyright claims and create contract claims against the AI company. Also, does it matter what kind of output the AI makes with the training? In other words, if the AI simply absorbs the art on plaintiff artists' websites, uses the information to make predictions about visual art in general, but is unwilling or unable to create outputs that copy the style of these artists, would that make a difference in deciding whether a copyright violation occurred? I would venture that such activity would not lead to copyright liability. However, the moment a generative AI company starts competing with the artists or journalists whose work was used to train its AI model, then the evaluation changes. The copying, assuming there was copying involved and proven, is no longer benign. There is a cost to the artists and journalists, and this cost is likely an important factor to determine liability for the generative AI company.

Would the answer depend on the type of art fed into the generative AI model? For example, another case has been filed by Sony Music, Universal Music, and Warner Records against two AI start-ups that produce songs based on simple text prompts.[27] If these AI companies can generate infinite new songs that sound like they were recorded by Elvis Presley, should the Presley estate have a right to stop them from doing so? The Sony plaintiffs claimed that music generating AI would "saturate the market with machine-generated content that will directly compete

---

27. *UMG Recordings, Inc. et al.* v. *Uncharted Labs, Inc. et al.* (24 Civ. 4777 (AKH) (S.D.N.Y. Aug. 22, 2024) and *UMG Recordings, Inc. et al.* v. *Suno, Inc. et al.* (1:24-cv-11611, U.S. District Court for the District of Massachusetts).

with, cheapen, and ultimately drown out the genuine sound recordings on which [the plaintiffs' services were] built."[28] The CEO of defendant AI company Suno, Mikey Shulman, defended his company by claiming its AI model was developed to create new music by understanding previously-recorded music, saying, "Our technology is transformative; it is designed to generate completely new outputs, not to memorize and regurgitate pre-existing content."[29]

But, unlike news companies, it is clear that the music distribution companies are less worried about direct copying than they are concerned over the possibility of creating new music in the style of your favorite recording artist without that artist's participation. If a fan could push a button and produce enough music in Adele's style to release as a new Adele album, many would do so, and some of the output might be worth saving or distributing. But, even if that activity does not copy any lines or direct musical quotes from Adele's published songbook, is the activity fair to Adele or her music distributor? This is an entirely new model for music production, and some music recording artists are embracing it, rather than fighting it. For example, the singer Grimes has developed a partnership between a music distributor and her generative AI platform that allows fans to produce music through a generative AI trained on Grimes's voice. Grimes and her AI do not claim any ownership of the generated songs, but Grimes will receive 50 percent of the royalties earned through this process.[30] Less tech-savvy singers and songwriters may turn to litigation to protect not only their music libraries, but the possibility that the world will be flooded with new music reflecting their brands without their permission or participation.[31]

The music generation business model varies the questions about whether training on publicly-available digital files would be considered to

---

28. *Id.*

29. Marc Tracy, *Major Record Labels Sue A.I. Music Generators: The lawsuits say that Udio and Suno trained their products on reams of copyrighted music*, N.Y. Times, Jun. 25, 2024.

30. Antonio Pequeño IV, *Grimes Helps Artists Distribute Songs Using Her AI Voice—If They Split Royalties. Here's How It Works,* Forbes, Jul. 12, 2023.

31. Song libraries can be a big business. In December 2021, Bruce Springsteen sold his entire recording and song catalogue to Sony for approximately $550,000,000. In 2020, Nobel Prize for Literature winner Bob Dylan sold his songbook for more than $300,000,000. Sony Music and Universal Music, who respectively purchased these catalogues, are plaintiff's in the suit against companies who trained generative AI on existing songs.

be fair use of copyrighted material. Where the generative AI model is only trained on popular music so that it can generate economically competitive music, how can the generative AI company justify its copying and use of currently existing digital files? Music and its business models may be a special case where copying files for the purpose of training AI models is *per se* copyright violation. An LLM that ingests and trains on news articles does not need to compete against the creator of those new articles. It can be training a customer-service chatbot or an Alexa-style interface to speak and understand English language. However, the music generating AI model may have only one significant function—to produce music that competes with its training materials.

At least one U.S. state is addressing this issue already. Without the universality of federal copyright law, Tennessee passed the Ensuring Likeness, Voice, and Image Security Act (ELVIS Act) in 2024 to penalize people from imitating the work of specific vocal artists.[32] Previous Tennessee law only protected a person's name, image, and likeness, but the ELVIS Act extends protection to a person's voice, which, by legal definition, includes both a person's actual voice and simulations of that voice, and the law prohibits commercial use of these personal representations. The Act creates a private right of action against anyone using a likeness or voice without permission and against anyone distributing tools that allows creation of such deepfakes. Given that many people have made a brand of their voice, or that their broader brand is characterized by the voice, this may be a special case.

In other words, singers and actors can be represented by their voice alone, and generating a person's voice to broadcast words that the person never spoke is a particularly intimate form of fraud. As humans, we have not yet reached the point of hearing a recognizable voice and easily dismissing that recording as an unauthorized fake. We tend to believe our ears. This is different from seeing a new work in a recognizable style of painting and understanding that the art was faked. A voice and a face are the most visceral and recognizable features of a person. This is the terror of a case like the school administrator in Maryland discussed in Final Note, who was the victim of deepfaked vocal files. Whether you are a public personality who makes money for your brand, or you are a private

32. ELVIS Act, signed on Mar. 21, 2024, *The N.Y. Times Co.* v. *Microsoft Corp.*, No. 23-cv-11195 (SHS) (OTW) (S.D.N.Y. Nov. 22, 2024).

person hoping to avoid the spotlight, a deepfake of your voice sounding words you never spoke can ruin your life.

As with so many technological advances of the past three decades,[33] one of the first generative AI applications was famously used to facilitate the development of pornography. These pornographic deepfakes targeted women by making videos that seemed to show a chosen deepfake subject involved in sexual activity. The videos and pictures are fraudulent. According to University of Pennsylvania researcher Sophie Maddocks, "Deepfakes are visual content created using AI technology, which anyone can access through apps and websites. The technology can use deep learning algorithms that are trained to remove clothes from images of women, and replace them with images of naked body parts. . . . Deepfakes really caught the public's attention in 2017, and two years later, in 2019, there were 14,678 deepfake videos online and 96% were pornographic. At the time all of these featured women, according to research from Deeptrace Labs."[34] Pornographic deepfakes are a category of the broadly-defined digital sexual abuse, where digital video or still-picture files are made of former girlfriends, online enemies, or celebrities and are released online.

Problems caused by porn deepfakes goes beyond reputations. The use of deepfakes and other generative AI technology in child pornography is

---

33. "From the printing press to instant cameras, from pay-per-view to VCRs, pornographers – both professional and private – have been among the quickest to jump on board with newly developed gadgets." Doug Gross, *In the tech world, porn quietly leads the way*, CNN, Apr. 23, 2010. In recent years, the online sex industry has provided high-earning business models to assist the introduction of digital streaming and bandwidth-reducing video tech, video chat, digital camera adoption, high-speed mobile internet, and digital advertising and marketing.

"With VCRs and cable TV, the adult industry drove a crucial wave of early adoption. . . . The ultimate in private access to pornography came with the internet. On the web, not even the check-out-counter guy at the video store has to know what you're up to." Kaveh Waddell, *How Porn Leads People to Upgrade Their Tech: The adult industry has a history of ushering in media revolutions*, The Atlantic, Jun. 7, 2016.

Next wave innovations like virtual and augmented reality, including lifelike AI-driven sex dolls and a full-body "cyber-sex" suit with thirty-six contact points, may be too expensive or niche-oriented to drive significant changes in the current technology market.

34. Hailey Reissman, *What Is Deepfake Porn and Why Is It Thriving in the Age of AI? In this Q&A, doctoral candidate Sophie Maddocks addresses the growing problem of image-based sexual abuse*, University of Pennsylvania Annenberg School for Communication, Jul. 13, 2023, https://www.asc.upenn.edu/news-events/news/what-deepfake-porn-and-why-it-thriving-age-ai#:~:text=These%20synthetic%20sexual%20images%20are,from%20non%2Dconsensually%20distributed%20images.

a tragic corner of internet and culture. In April 2023, a Quebec man was sentenced to more than three years in prison for using artificial intelligence to produce synthetic videos of child pornography. The judge in the case noted how a single photograph of a child could be used to project AI-enabled abusive digital files, and that "the creation of the new images of sexual abuse encourages the market for child pornography, which craves novelty, and puts children at risk by 'fuelling fantasies that incite sexual offences against children.'"[35] This activity is likely criminalized in the United States by many state laws and by the federal Protection of Children Against Sexual Exploitation Act of 1977.

As a legal problem, sexually-abusive deepfakes clearly fall in the category of fraud, but their insidious effects stretch far beyond regular fakery. Victims of this crime may not be able to prove a negative; it may be impossible to establish that the victim did not appear in the digital video. In addition, the simple act of viewing a deepfake sexual abuse video can affect the viewer's impression and understanding of the victim with no means to recover a pre-viewing state of mind, so damages to the victim may be severe, but will always be impossible to accurately quantify. How does a victim demonstrate legally-recognizable damages? Several U.S. states now have laws that prohibit distributing sexually-oriented content, which can include deepfakes.[36] The United Kingdom also has a law against distributing such abusive material.[37] These laws criminalize the specific behavior of digital sexual abuse and make it easier for women in these jurisdictions to both pursue their attackers and to remove the offensive digital content from publicly accessible spaces.

Lawyers have a vested interest in policy and politics, and deepfakes are invading that arena as well. In 2016, an established flood of false information was unleashed on the American public in advance of the Presidential election. Much of this fraud was created by Russians for the sole purpose of dividing the electorate and increasing anger of members of both parties

35. Jacob Serebrin, *Quebec man who created synthetic, AI-generated child pornography sentenced to prison: Judge calls use of deepfake technology to create images 'chilling'*, The Canadian Press, Apr. 26, 2023.

36. *See*, for example, New York Civil Rights Law § 52-C, Florida Statutes § 836.13, Georgia Code § 16-11-90, California Civil Code § 1708.86, Texas Penal Code § 21.165, and Minnesota Statutes § 604.32.

37. U.K.'s Online Safety Act.

at people on the other side.[38] As deepfake technology has both improved and become easier to use with standard computing equipment, election experts were worried about its use as a new form of political fraud. A faked version of Joe Biden's voice was used in automatically generated robocalls to discourage Democrats from taking part in the primary.[39] Deepfakes spreading across the web can be countered quickly by attentive candidate teams, but those distributed through direct telephone calls are more difficult to catch and fight. According to Josh Lawson, director of AI and Democracy at the Aspen Institute, "The concern . . . is not the big, bad deepfake of somebody at the top of the ticket, where all kinds of national press is going to be out there to verify it. It's about your local mayor's race. It's about misinformation that's harder and harder for local journalists to tackle, when those local journalists exist at all."[40] Election deepfakery has reached the attention of legislators as well, with acts being introduced to address the problem.[41] Election deepfakes are a very specific societal problem created by generative AI, and this problem has clear negative implications for the effective management of a democracy, so this is an area where laws may pass to address the issue.

One of the clearest legal concerns about deepfakes is how they will affect court proceedings when offered as evidence, or conversely, when a genuine article of evidence is challenged as an AI-generated falsehood. New technology that allows sophisticated falsifying of evidence that can be emotionally resonant and difficult to disprove is a perfect storm to sew confusion in the courtroom. A new fraudulent tool improves the ability of bad actors to defraud the court and win their litigation with realistic faked evidence, or conversely to drag out litigation and throw doubt on virtuous cases by constantly questioning the validity of evidence. However, whether or not these underhanded strategies will be successful, the possibility of their assertion will affect other litigation. Academic experts on courtroom management predict an explosion of costs as people submitting

---

38. U.S. Department of Justice, *Report On The Investigation Into Russian Interference In The 2016 Presidential Election*, https://www.justice.gov/archives/sco/file/1373816/dl.

39. Zachary B. Wolf, *The deepfake era of US politics is upon us*, CNN, Jan. 23, 2024.

40. Shannon Bond, *AI fakes raise election risks as lawmakers and tech companies scramble to catch up*, NPR, Feb. 8, 2024.

41. *See*, for example, H.R.6088—Deepfakes in Federal Elections Prohibition Act, 116th Congress (2019–2020); H.R.5586—DEEPFAKES Accountability Act, 118th Congress (2023–2024).

AI-generated content and people accused of doing so will be forced to hire expensive technical and forensic evidentiary experts to resolve the validity of the proposed evidence. "This will obviously serve to increase the cost of litigation in an already unaffordable justice system. . . . These hurdles can be expected to cause a crisis for criminal defendants and public defenders who simply cannot afford the kinds of expensive experts that will be needed to mount a proper defense. More appeals based on a claim of ineffective assistance of counsel may also result."[42] In court, the cost of process can be a deciding factor in which cases are litigated to completion and which are settled. Generative AI deepfakes could lead to more settlements as cost rise for proving cases affected by the threat of competently falsified evidence.

Are new courtroom rules needed to address AI-generated deepfakes? Judge Paul Grimm and Professor Maura Grossman have formally proposed a new Federal Rule of Evidence[43] stating, "If a party challenging the authenticity of computer-generated or other electronic evidence demonstrates to the court that it is more likely than not either fabricated, or altered in whole or in part, the evidence is admissible only if the proponent demonstrates that its probative value outweighs its prejudicial effect on the party challenging the evidence."[44] They proposed this rule so that the party challenging the evidence as an AI-generated fake must bear the burden of establishing to the court that the evidence (or material part of the evidence) was more likely than not a fake. If the challenging party meets this burden, then the party offering the evidence has the burden of establishing the evidence is factual and that its probative value in the case outweighs its prejudicial value on the opposing party. This new rule has not been adopted, but may be in the future as courts have more experience with generative AI-based challenges to evidence.

---

42. Maura R. Grossman, Hon. Paul W. Grimm (Ret.), Daniel G. Brown & Molly (Yiming) Xu, *The GPTJudge: Justice in a Generative AI World*, 23(1) Duke L. & Tech. Rev. 1–34 (2023).

43. Proposed Federal Rule of Evidence 901(c): Potentially Fabricated or Altered Electronic Evidence, introduced by Grimm and Grossman at the April 2024 meeting of the Advisory Committee on Evidence Rules.

44. Benj Edwards, *Deepfakes in the courtroom: US judicial panel debates new AI evidence rules: Panel of eight judges confronts deep-faking AI tech that may undermine legal trials*, Ars Technica, Apr. 4, 2024.

A different law professor has proposed a version of evidentiary changes which would have the opposite effect. Professor Rebecca A. Delfino of Loyola Law School in Los Angeles has proposed that the party offering evidence be charged with the burden of proving to the judge that the evidence is valid.[45] She proposes that the judge should hold a hearing on authenticity, and if the evidence is admitted, the judge must issue a statement to the jury not to doubt the evidence simply because of the existence of deepfakes. She argues that this change is needed for evidentiary rules because "given the speed of technological advances in the field of AI and deepfakes and the disruption to the truth-finding process they pose, the incremental approach to the issues of admissibility found in the common law is not an effective way to deal with deepfakes."[46] And a 2020 proposal suggested an evidentiary requirement for corroboration of video evidence to address the AI deepfake threat.[47] So even at this early stage of generative AI infiltration into society, legal and academic opinions differ on the best method to address the impact of deepfakes on evidence in litigation.

At this writing, such concerns are premature. U.S. courts have not yet been swamped with generative AI fakes offered as evidence or lawyers accusing their opponents of submitting deepfakes. Some writers argue that deepfakes will never be a serious problem and AI misinformation is the threat "that always seems just around the corner."[48] As noted above, such a false submission of evidence could lead to serious disciplinary action against the submitting attorney. Respect for the rules may be the

---

45. Rebecca A. Delfino, *Deepfakes on Trial: A Call To Expand the Trial Judge's Gatekeeping Role To Protect Legal Proceedings from Technological Fakery*, 74(2) Hastings L.J. 293 (2023).

46. *Id.* at p. 332.

47. John P. LaMonaga, *A Break From Reality: Modernizing Authentication Standards for Digital Video Evidence in the Era of Deepfakes* 69(6) Am. U. L. Rev. 1945, 1984 (2020), proposing new Federal Rule of Evidence 901(b)(11) which would state: "Before a court admits photographic evidence under this rule, a party may request a hearing requiring the proponent to corroborate the source of information by additional sources."

48. Peter Carlyon, *Deepfakes Aren't the Disinformation Threat They're Made Out to Be*, RAND Research & Commentary Blog, Dec. 19, 2023. Carlyon looks at deepfakes through the broader context of misinformation and finds, "catastrophizing deepfakes misunderstands the message-driven way they are often used and misrepresents disinformation. . . . concerns over generative AI too often stray into alarmism, of hypothetical dystopias where fact is indistinguishable from fiction. That has distorted the picture."

legal system's saving grace in facing a potential landslide of machine-generated fakery. Courts, bar associations, and legislatures will continue to address this issue in coming years.

Less than two years into the generative AI revolution, the United States has seen both legislative and regulatory[49] activity aimed at managing the cultural and legal impact of this new technology. In the Spring and Summer of 2024, the California legislature was considering seven separate AI bills which passed their chamber of origin, many addressing generative AI concerns. For example, California enacted a generative AI bill aimed at protecting performers in the entertainment industry. Some have been enacted at the time of this writing. While the rights of voice actors were recognized as threatened by generative AI deepfakes, companies have argued that performers whose body movements are used to animate video games are acting as motion capture and not as protected performances. Taking its lead from positions taken by actors in their 2023 strike against production studios, the California law protects performers against "digital replicas" of their performances being used without the performer's express written permission. The California legislature provided that "digital replica" means "a computer-generated, highly realistic electronic representation that is readily identifiable as the voice or visual likeness of an individual that is embodied in a sound recording, image, audiovisual work, or transmission in which the actual individual either did not actually perform or appear, or the actual individual did perform or appear, but the fundamental character of the performance or appearance has been materially altered, except as prescribed."[50] This new law would mean California performers could no longer be forced to relinquish their rights to their likeness, a common recent practice in the entertainment industry as more of the business, product, and production process becomes digitized. This group of bills also included legislation prohibiting use of a dead person's voice, body, or face likeness and one that required large online platforms like Meta's services to remove deepfakes related to elections.

In 2023, the Biden Administration issued an executive order on trustworthy artificial intelligence that affected nearly every unit of the U.S.

---

49. Chapter 4 discussed actions taken by both the U.S. Copyright Office and the U.S. Patent and Trademark Office to address questions of whether or how generative AI models or their owners could be credited with government protection for AI-generated content.

50. AB 2602 which amends § 3344.1 of the California Civil Code (effective Jan. 1, 2025).

executive branch.[51] The order covers dozens of topics, handing assignments to National Institute of Standards and Technology (NIST) and other agencies, demanding AI governance plans across the federal government, and proposing methods to avoid deepfakes in government policy statements. It starts with developing guidelines, standards, and best practices for AI safety and security, moves to reducing AI risks in regard to weapon systems, and includes preventing malicious use of federal data for AI training. Interestingly, it also proposes watermarking generative AI output so that deepfakes may be distinguished from pictures, audio, and video files taken from reality. While the government claims to have received an assurance to work on such programs with "15 leading companies,"[52] progress on this topic has not been announced. This was not the first U.S. Executive Order on AI,[53] but it was the first to directly address the perceived dangers of generative AI for the government.

The European Union Artificial Intelligence Act (EU AI Act) was written and passed before popular imagination was triggered by generative AI. However, its comprehensive framework for developing and using AI will also have an effect on AI generating content for consumers in the European Union. The one prohibited activity under the EU AI Act that relates to this category of functionality is cognitive behavioral manipulation of people or specific vulnerable groups, which could be managed by a chatbot or other LLM. Since such activities are strictly forbidden in Europe under this Act, companies developing and utilizing AI chatbots will need to train and monitor them very carefully to avoid accusations of manipulation. This will be especially true for companies selling toys or in-game applications for children because AI interactions with vulnerable users are automatically suspect under this Act. Further, chatbots used in marketing, rather than customer service or employee benefits, run the risk of being ruled manipulative simply by providing opportunities to purchase goods or services through the AI-driven chat function. High-risk (as opposed

51. *Executive Order on the Safe, Secure, and Trustworthy Development and Use of Artificial Intelligence*, Briefing Room, The White House, Oct. 30, 2023.

52. *FACT SHEET: President Biden Issues Executive Order on Safe, Secure, and Trustworthy Artificial Intelligence*, Briefing Room, The White House, Oct. 30, 2023.

53. For example, the Trump Administration released an executive order on AI (Feb. 12, 2019). This order made a feint toward pushing development of trustworthy AI standards, but was primarily an AI investment announcement.

to "prohibited") AI models tend to fall into this classification from the general area of society they fall into, rather than their functionality. So high-risk generative AI may possibly be used for critical infrastructure management, education and vocational training, employment and hiring, law enforcement, migration, or the courts, and therefore have significant regulatory obligations under the EU AI Act.

Generative AI has been the gateway drug for public recognition of how AI can affect people's lives. This functional category of AI is drawing billions of development dollars and will only grow in effectiveness and impact. Lawyers are affected by generative AI, but unlikely to be replaced by it. The legal issues generative AI raises in intellectual property and creative business will ripple through the law and economy for generations. But generative AI is not the only AI functionality that matters to the world, and lawyers who stop their focus here will be missing important technological advancements and legal issue.

CHAPTER 6

Personal Identification AI

Dozens of databases around the world probably hold biometric readings identifying you. The FBI may have your fingerprints. The Chinese government could have your face recorded. Disney may have your hand geometry. Your bank may have your voiceprint on file. Your phone provider not only takes your fingerprint or faceprint (probably stored on your phone), but it also likely has a record of how you hold your phone and enter keystrokes. Your television and your gaming system may know exactly who is in the room with them while they are on and how long your eyes stay focused on the screen. If DNA databases like Ancestry.com hold your cousin's biometric samples, then they know things about you too.

I have intentionally given my fingerprints and face scan to the U.S. Transportation Security Administration. Just time yourself in lines between the parking lot and the airplanes at Hartsfield-Jackson Atlanta International Airport without the reduced security delay that a biometric signature allows. The prescreened time savings is significant and can be crucial after fighting Atlanta traffic. I have given my fingerprints to the state bar association, and I'm not sure what broader databases also contain these samples.

All of these systems apply some form of AI or otherwise sophisticated algorithm, often in multiple places. They would not be effective without AI. This book has often noted that AI tends to be particularly good at tasks of sorting, grouping, differentiating, predicting, and identifying. All of these tasks are important in biometric identification. An AI model may break a biometric reading—retinal scan, fingerprint, face geometry, and voice recording—into component parts that can be measured and compared.

A different AI could be used differentiate among these databases. A third AI (or possibly the same AI model) may be applied to make a comparison between a newly-entered biometric reading and those already recorded in the database to find some level of match. This is the method illustrated in countless procedural law enforcement television shows where a fingerprint is found at the crime scene and then run through the FBI fingerprint database to discover an eight-point match, or even better, a sixteen-point match, with a fingerprint taken from a person previously convicted of a similar crime. The television police now have a suspect for their crime.

But what if just being in a certain location is considered to be a crime? We live in a world of ubiquitous surveillance. Most cities have cameras at every major intersection and scores of public and private cameras throughout the landscape. We carry small tracking devices called smartphones with us wherever we go, and journalist Kashmir Hill showed us that our personal vehicles are geo-sensing computers that constantly track our locations and behaviors, even if we specifically ask them not to.[1] We bring cameras and microphones into our homes and then connect them to the wider world via the internet. The internet itself, through social media, digital shopping, electronic news programs, and self-publishing, carves windows into our feelings, priorities, and preferences that simply did not exist prior to 1990.[2]

Changes in technology have been adopted by law enforcement for at least 150 years and have shifted the nature of the relationship between government and citizenry. When photography was introduced, mug shots made identifications much easier. Electric lights facilitated searches. Replacing horses with automobiles not only made long-distance travel easier, but the automobile became a new stage to conduct personal searches. The application of inexpensive listening and recording devices allowed deeper intrusion into the lives of witnesses and suspects. Computers, lapel cameras, drones, Stingray devices, military hardware, and

---

1. And those companies frequently sell all of that information to our insurance carriers. So if you thought you were avoiding surveillance by refusing to install an insurance tracking device in your car or truck, the insurance company may have received your driving information anyway from the vehicle's manufacturer. Kashmir Hill, *Automakers Are Sharing Consumers' Driving Behavior With Insurance Companies: LexisNexis, which generates consumer risk profiles for the insurers, knew about every trip G.M. drivers had taken in their cars, including when they sped, braked too hard or accelerated rapidly*, N.Y. Times, Mar. 11, 2024.

2. Alexandra Rengel, *Privacy-Invading Technologies and Recommendations for Designing a Better Future for Privacy Rights*, 8 Intercultural Hum. Rts. L. Rev. 177, 178–79 (2013).

AI databases containing boundless biometric identification markers all change the nature of policing. For a century, U.S. courts have noted how technological changes may affect constitutional rights. In his famous 1928 Supreme Court dissent, Justice Brandeis noted the march of technology and its effect on Fourth Amendment rights, writing:

> Discovery and invention have made it possible for the Government, by means far more effective than stretching upon the rack, to obtain disclosure in court of what is whispered in the closet. . . . Moreover, 'in the application of a Constitution, our contemplation cannot be only of what has been but of what may be.' The progress of science in furnishing the Government with means of espionage is not likely to stop with wiretapping. Ways may someday be developed by which the Government, without removing papers from secret drawers, can reproduce them in court, and by which it will be enabled to expose to a jury the most intimate occurrences of the home.[3]

In 2001, a unanimous Supreme Court found that the "advance of technology" clearly affects "the degree of privacy secured by citizens under the Fourth Amendment," observing that human flight had made previously private spaces more vulnerable to observation.[4]

Surveillance has increased because information capturing technology improved in a spectacular fashion. Many police departments now use Stingray devices that mimic cell towers to capture the content of phone calls. Others operate self-controlled drones as first responders. Networks of cameras on houses, banks, intersections, and street corners are available for law enforcement to peruse at will. Our vehicles and personal devices not only show police where we are, but they will also do so over time, so law enforcement can track where you were at key moments in the past and where you can be found right now. In 2021, the U.S. Government Accounting Office surveyed forty-two federal agencies that employ law enforcement officers about their use of facial recognition technology, and twenty reported owning such systems or using systems owned by others.[5]

---

3. *Olmstead* v. *United States*, 277 U.S. 438, 474 (1928).

4. *Kyllo* v. *United States*, 533 U.S. 27, 33–34 (2001).

5. *Facial Recognition Technology: Federal Law Enforcement Agencies Should Better Assess Privacy and Other Risks*, Government Accounting Office, GAO-21-518, Jun. 3, 2021.

The application of AI has advanced to a point where computers assist in matching license plates, fingerprints, or other identifying information captured by law enforcement. So now the massive databases of surveillance cameras are knitted together and their results can be compared against other databases. Police can highlight the license plate of a truck in one part of town, and run AI programs to see where that same license number appeared at different times in different locations. The same can also be true for facial recognition AI, tracking the appearance of one face on multiple different cameras in different locations. Some security AI is used to highlight when problems arise. Police employ recording tools to pinpoint when and where gunshots occur in a city, and the complementary AI sends out warning and provides information related to the shot. Similarly, AI can be set to search for specific people's faces on banks of cameras (like those at a ballpark) and for when the AI spots human behavior that likely indicates violence, or even predicts potential violence in the near future.[6]

Since the attacks on the World Trade Center in 2001, Americans have been reluctant to place limits on policing. But that dynamic has changed as the Black Lives Matter movement brought law enforcement overreach into the national conversation. However, states still seem wary of blocking police departments from gathering every possible tool available for catching criminals and pursuing potential terror suspects. This has led to a militarization of U.S. police forces and a nearly unchecked ability for law enforcement to adopt powerful new surveillance technologies without preliminary public policy discussion of advisability or even constitutionality. Lawyers should be asking whether the intrusiveness of these technologies and the AI that enables them violates the U.S. Constitutional Fourth Amendment protections against unlawful search and seizure.

---

6. "Over the past 15 years, the country's largest police departments . . . started thinking about ways of using artificial intelligence to not just analyze crime but predict it. They often turned to data analytics companies such as PredPol and Palantir, which create software that law enforcement can use to forecast crime. Predictive policing tools are built by feeding data—such as crime reports, arrest records and license plate images—to an algorithm, which is trained to look for patterns to predict where and when a certain type of crime will occur in the future." Prashnu Verma, *The never-ending quest to predict crime using AI: The practice has a long history of skewing police toward communities of color. But that hasn't stopped researchers from building crime-predicting tools*, The Washington Post, Jul. 15, 2022.

For example, facial recognition databases are continuously growing and the technology is engaged regularly by law enforcement agencies at all levels across the country.[7] Facial recognition software can be an unconstitutional impingement of search and seizure limitations for people to whom it is applied without suspicion of criminal activity, but the same software can be a constitutionally allowable method of identifying a suspect who was recorded on video committing a crime. Therefore, like nearly all tools, the technology itself is not constitutionally suspect, but certain applications of the technology may be. The same person can use a knife to carve a roast for dinner or to murder a neighbor; one application is a benefit and the other a crime. Similarly, the existence of AI tools or the simple fact that law enforcement agencies are using them may not be unlawful, but unconstitutional uses of the tool should be stopped.

An excellent way to address this problem is to require police to secure a warrant before applying biometric AI systems to identify people in pictures and videos. Asking law enforcement to secure a warrant before gathering certain evidence means that, in practice, the police seeking the evidence would need to show that they had a reasonable basis to suspect that the targeted person committed a crime and a reasonable basis to believe that the evidence sought would provide information leading to an arrest.[8] This showing is called "probable cause" to establish that a reasonable person would believe that a crime was in the process of being committed, had been committed, or was going to be committed.[9] U.S. courts do not currently require a warrant to access AI identification models, despite

7. *See*, Kristin Finklea, Laurie A. Harris, Abigail F. Kolker & John F. Sargent Jr., *Federal Law Enforcement Use of Facial Recognition Technology*, R46586, Congressional Research Service Report, Oct. 27, 2020, https://fas.org/sgp/crs/misc/R46586.pdf, stating, "The [FBI] operates two programs that support the use of the technology: (1) the Next Generation Identification–Interstate Photo System (NGI-IPS), largely supporting state and local law enforcement; and (2) the Facial Analysis, Comparison, and Evaluation (FACE) Services Unit, supporting FBI investigations. NGI-IPS contains criminal mugshots, and the system allows authorized law enforcement users (primarily state and local) to search the database for potential investigative leads. The FACE Services Unit supports FBI investigations by searching *probe* photos of unknown persons against faces in NGI-IPS and other federal and state facial recognition systems authorized for FBI use."

8. *See*, U.S. Constitution, Article 4, stating, "The right of the people to be secure in their persons, houses, papers, and effects, against unreasonable searches and seizures, shall not be violated, and no Warrants shall issue, but upon probable cause, supported by Oath or affirmation, and particularly describing the place to be searched, and the persons or things to be seized."

9. *Id.*

the Constitutional issues. To change this caselaw without legislation, there would need to be a case where someone was convicted due to unwarranted access to an AI matching database. Such cases have not worked their way far enough into the courts yet to receive a definitive review.

A warrant, where required, is easy to apply in practice and its requirement for this purpose appropriately modernizes the operation of the Fourth Amendment to the U.S. Constitution to address technologies that could not have been conceived by our founders but affect the rights of citizens every day. Furthermore, a warrant requirement for law enforcement application of this technology falls in line with recent U.S. Supreme Court cases where electronic surveillance has been found unconstitutionally intrusive.[10] Facial recognition software can be used for purposes that some people would find violative of Fourth Amendment protections against improper searches, like identifying people peacefully gathering to assert their political rights. The Supreme Court has held that U.S. citizens have a right to remain anonymous in their political activity, stating, "[i]nviolability of privacy in group association may in many circumstances be indispensable to preservation of freedom of association, particularly where a group espouses dissident beliefs."[11] Indiscriminate or politically targeted applications of facial-recognition AI searching politically motivated crowds could violate that right. The ability to focus on face after face, revealing name after name, should be examined for constitutionality unless the government's motivation clearly falls within permissible search parameters.

Consider an American crowd lawfully demonstrating against the police force itself—this could be because the police are enforcing restrictive gun laws or because of a discriminatory use of force. We have seen that groups across the political spectrum could have a reason to protest law enforcement policies or behaviors. For centuries, demonstrators have relied on the anonymity of a crowd to protect them from government retaliation. Technology makes today's world different. Police can scan the crowd using cameras in place for traffic control or business security, cameras worn by uniformed officers or wielded by members of the crowd

10. *See*, for example, *Carpenter* v. *United States*, 138 S. Ct. 2206, 2223 (2018); *Riley* v. *California*, 573 U.S. 373, 403 (2014).

11. *NAACP* v. *State of Alabama ex rel. Patterson*, 357 U.S. 449, 462 (1958) (stating that "compelled disclosure of affiliation with groups engaged in advocacy may constitute" serious restraints on First Amendment rights).

itself, or even cameras flown over the demonstration on drones.[12] They can save and store the pictures to study at a later date and then run them against identifying AI. The Constitution would allow law enforcement to use face-matching AI against faces of people who clearly threw flaming objects into stores, but the Constitution seems to require that same police unit needs a warrant—with the concomitant showing of probable cause that the targeted people committed crime—to search for names of people far away from criminal action.

In addition, despite its level of detail, biometric identification is not perfect and can lead to false identifications which can ruin people's lives. For example, following the 2004 bombing of the train station in Madrid, Brandon Mayfield, a lawyer in Portland, Oregon, was wrongfully arrested and held for weeks by the FBI based on a mistaken fingerprint match at the FBI labs. The FBI had mistakenly matched Mayfield's fingerprints with prints on a plastic bag found near the scene of the attacks that killed 191 people. Mayfield, an Islamic convert married to an Egyptian woman, felt he was being watched before he was arrested and held by the government. He was released after the FBI realized its mistake. A government report later found, "fingerprint examiners were not asked to make an objective determination of whether the prints matched; instead, they were asked whether they could confirm a match made by another FBI technician."[13]

The identifying AI technology itself changes the level of intrusion into the lives of private citizens by their government, which is a primary space for legal work and judicial commentary. This functional category of AI raises Constitutional issues about how much government intrusion into peoples' lives we will allow. In describing the negative effects of Global Positioning System (GPS) monitoring by police, U.S. Supreme Court Justice Sotomayor wrote:

> Awareness that the Government may be watching chills associational and expressive freedoms. And the Government's unrestrained power to assemble data that reveal private aspects of

12. Aaron Holmes, *How police are using technology like drones and facial recognition to monitor protests and track people across the US*, Business Insider, Jun. 1, 2020, https://www.businessinsider.com/how-police-use-tech-facial-recognition-ai-drones-2019-10#cell-tower-simulators-or-stingrays-5.

13. Melissa Block, *FBI Faulted in Madrid Bombing Fingerprint Error*, National Public Radio, Jan. 6, 2006.

> identity is susceptible to abuse. The net result is that GPS monitoring—by making available at a relatively low cost such a substantial quantum of intimate information about any person whom the Government, in its unfettered discretion, chooses to track—may 'alter the relationship between citizen and government in a way that is inimical to democratic society.'[14]

The Supreme Court was examining the GPS tracking technology because it allowed an entirely new ability for the police to intrude on private lives. In earlier eras, no one would expect the police to be able to track a citizen's privacy movements every minute twenty-four hours a day for a month. The task was simply too expensive, too difficult, and too deep a drain on manpower. But with a GPS tracker, law enforcement was checking a single person's locations every thirty seconds for more than a month. So the Supreme Court held that a warrant was needed to use such intrusive technology.[15] The same should be true for AI-driven identification technology where law enforcement captures a person's biometric readings and then runs the biometric through databases with AI matching.

People have a certain expectation of obscurity even in a public place, and the AI identification allows police intrusion so far beyond what was available in earlier eras that it destroys those expectations. In a recent case, Chief Justice John Roberts wrote, "[a] person does not surrender all Fourth Amendment protection by venturing into the public sphere."[16] To the contrary, "what [one] seeks to preserve as private, even in an area accessible to the public, may be constitutionally protected."[17] Roberts then noted how technology has changed the calculation of what should be considered a reasonable expectation of privacy even in the public sphere, writing, "[i]n the past, attempts to reconstruct a person's movements were limited by a dearth of records and the frailties of recollection . . . [with current technology] police need not even know in advance whether they want to follow a particular individual, or when."[18]

---

14. *United States* v. *Jones*, 565 U.S. 400, 416 (2012) (Sotomayor, J., concurring) (quoting *United States* v. *Cuevas-Perez*, 640 F.3d 272, 285 (7th Cir. 2011) (Flaum, J., concurring)).

15. *United States* v. *Jones*, 565 U.S. 400 (2012), majority opinion.

16. *Carpenter* v. *United States*, 138 S. Ct. 2206, at 2217 (2018).

17. *Id.*

18. *Id.* at 2218.

The cameras, the AI management of the databases, and the AI identification of a person of interest does all the work that the police could never do even a few years ago. The new technology of AI facial recognition software, like the GPS data or cell phone triangulation information pulled from a citizen's smartphone, holds the potential to reveal aspects of a person's life that are both private and constitutionally-protected. The Supreme Court examined cell phone technology and found, "Here the progress of science has afforded law enforcement a powerful new tool to carry out its important responsibilities. At the same time, this tool risks Government encroachment of the sort the Framers, 'after consulting the lessons of history,' drafted the Fourth Amendment to prevent."[19] This is also an accurate description of AI-enabled biometric identification technology. The government can use it as a powerful new tool for important law enforcement functions, but it also can enable the kind of government intrusion that the founders warned about, such as indiscriminate or coercive searches.

While warrants are not universally required for law enforcement to access the AI-driven facial recognition databases, some states have otherwise recognized the risks involved with allowing police to use such powerful AI tools without a procedural check. Multiple state and local laws place restrictions on police using AI identifying facial recognition technology.[20] For example, both Massachusetts and Utah require law enforcement to submit written requests to the state agency maintaining the database to conduct a facial recognition search.[21] Virginia passed multiple limitations on government use of facial recognition technology, with a rule stating, "A local law-enforcement agency shall not (i) use facial recognition technology for tracking the movements of an identified individual in a public space in real time; (ii) create a database of images using a live video

---

19. *Id.* at 2223 (citing *United States* v. *Di Re*, 332 U.S. 581, 595 (1948)).

20. Cities that have banned law enforcement access to facial recognition technology: San Francisco, Oakland, and Berkley in California; Boston, Brookline, Cambridge, Northampton, Easthampton, and Somerville in Massachusetts; Jackson, Mississippi; King County, Washington; Madison, Wisconsin; New Orleans, Louisiana; Minneapolis, Minnesota; Portland, Maine; and Portland, Oregon.

21. Mass. Gen. Laws 6 § 220: Facial recognition searches; requests; valid purposes; documentation; reporting; exceptions and Utah Code § 77-23e-103: Government use of facial recognition system with image database - Restrictions - Process - Disclosure.

feed for the purpose of using facial recognition technology; or (iii) enroll a comparison image in a commercial image repository of a facial recognition technology service provider except pursuant to an authorized use."[22] The statute allows up to fourteen authorized uses, including helping identify a person when there is a reasonable suspicion that person has committed a crime, helping identify a missing person or crime victim, or helping identify a person who is lawfully detained. Several states have different types of limitations on AI facial recognition use in law enforcement. For example, Minnesota bans the use of facial recognition technology in conjunction with police body cameras or drones.[23] Alabama law prohibits making facial recognition identification the sole basis of an arrest.[24] The State of Washington takes my preferred route toward protection of citizens' rights by requiring that police secure a warrant or court order prior to using facial recognition technology.[25]

Some cities have promoted the use of AI facial identification, only to find the results problematic.[26] Conversely, after a wave of these state and local laws was passed in the early 2020s restricting police use of AI facial recognition, there has been a subsequent loosening of these restrictions as some locations find that the convenience the technology offers law enforcement may outweigh the restraints on personal privacy. In 2022, *Reuters* reported, "Facial recognition is making a comeback in the United States as bans to thwart the technology and curb racial bias in policing come under threat amid a surge in crime and increased lobbying from developers. Virginia . . . will eliminate its prohibition on local police use of facial recognition a year after approving it, and California and the city of New Orleans as soon as this month could be next to hit the undo button."[27]

---

22. Code of Virginia § 15.2-1723.2: Facial recognition technology; approval; penalty (effective date Jul. 1, 2026).

23. Minnesota Statutes § 44-15-106.

24. Code of Alabama § 15-10-111.

25. Revised Code of Washington § 43.386.080.

26. Alfred Ng, *'Wholly ineffective and pretty obviously racist': Inside New Orleans' struggle with facial-recognition policing: Records obtained and analyzed by POLITICO reveal the practice failed to identify suspects a majority of the time and is disproportionately used against Black people*, POLITICO, Oct. 31, 2023.

27. Paresh Dave, *Focus: U.S. cities are backing off banning facial recognition as crime rises*, REUTERS, May 12, 2022. *See also*, "Five municipal bans on police and government use passed in 2021, but none in 2022 or in 2023 so far, according to a database from the digital rights group Fight for

Other jurisdictions restrict private use of the same AI facial identification systems. Most significantly, the Illinois Biometric Identification Privacy Act (BIPA)[28] limits private companies from taking biometric readings, which can include face geometry, for the purposes of identifying a person without that person's permission. This Act has been in force since 2008, and the combination of a private right of action with a provision for statutory damages has led to hundreds of lawsuits against private companies who are using biometric identifiers for employees or consumers. AI-based biometric systems measuring all types of biological readings have been challenged under BIPA. Other states have similar biometric identification restrictions, but they enforce those laws through the state attorney general's office.[29] New York City has passed an Act requiring disclosure of facial recognition technology to customers with clear and conspicuous sign, and prohibiting NYC businesses that "collect, retain, convert, store, or share biometric identifier information of customers" from profiting off the information.[30] Portland, Oregon prohibits companies from using facial recognition systems in places of public accommodation.[31] And for about a year, no resident or corporation in the City of Baltimore could use facial recognition technology or any information obtained from one of these systems. The Baltimore rule expired and has not been reinstated to date.[32]

Many of the arguments in favor of these limitations highlighted the problem of AI bias in biometric systems. In chapter 4, we covered the many fundamental logic and training flaws that could lead to bias in AI decision-making, but AI biometric identification systems have their own bias issues. The technology has a much higher accuracy rate for white men, then it has for women and dark-skinned people. The American Civil Liberties Union (ACLU) reports, "Studies show that facial recognition

---

the Future. Some local bans have even been partially repealed, and today, few seriously believe that a federal ban on police use of face recognition could pass in the foreseeable future." Tate Ryan-Mosley, *The movement to limit face recognition tech might finally get a win: A Massachusetts bill restricting police use could set the standard for how the technology is regulated in America. If it fails, it'll be a blow to a once-promising movement*, MIT Tech. Rev., Jul. 20, 2023.

28. Illinois Compiled Statutes Chapter 740 § 14/15.

29. Revised Code of Washington Chapter 19.375 and Texas Code Title 11, Subtitle A, Chapter 503.

30. NYC Admin. Code § 22-1202.

31. Portland City Code Chapter 34.10.

32. Baltimore City Ordinance 21-038.

technology is biased. The error rate for light-skinned men is 0.8%, compared to 34.7% for darker-skinned women, according to a 2018 study.... A 2019 test by the federal government concluded the technology works best on middle-age white men. The accuracy rates weren't impressive for people of color, women, children, and elderly individuals."[33] So aside from human bias that might slip into AI facial recognition training, and the various logical problems discussed earlier, we are faced with what seems to be a reading and recognition issue: AI facial identification systems have so far found it easier to correctly identify middle-aged white men then other people.

And these biases have consequences. There are numerous anecdotes of darker-skinned people who were mis-identified as criminals by AI facial matching systems and subsequently arrested by law enforcement. *Scientific American* ran a study on use of these AI systems and stated, "We found that law enforcement agencies that use automated facial recognition disproportionately arrest Black people. We believe this results from factors that include the lack of Black faces in the algorithms' training data sets, a belief that these programs are infallible and a tendency of officers' own biases to magnify these issues."[34] This AI study found that "Police typically use this technology in three ways: in-field queries to identify stopped or arrested persons, searches of video footage or real-time scans of people passing surveillance cameras. The police upload an image, and in a matter of seconds the software compares the image to numerous photos to generate a lineup of potential suspects."[35] The privacy violations in this activity becomes more frightening when you understand that the AI system itself has biases and inaccuracies. Many studies have dug into the reasons and consequences of facial recognition system bias,[36] and there

33. Rachel Fergus, *Biased Technology: The Automated Discrimination of Facial Recognition: Facial recognition remains largely ungoverned - and dangerous - in Minnesota*, ACLU of Minnesota, Feb. 29, 2024, quoting in the second sentence, a study titled "Gender Shades" by Joy Buolamwini and Timnit Gebru, published by MIT Media Lab.

34. Thaddeus L. Johnson & Natasha N. Johnson, *Police Facial Recognition Technology Can't Tell Black People Apart: AI-powered facial recognition will lead to increased racial profiling*, Scientific American, May 18, 2023.

35. *Id.*

36. For example, for a deep dive of facial recognition system bias, see Dr David Leslie, *Understanding bias in facial recognition technologies: An explainer*, The Alan Turing Institute, 2020, which states, "Harms arising from the deeply entrenched historical patterns of discrimination that are

will undoubtedly be more, but accept that such biases exist, they have been shown to hurt individuals and the administration of justice, and they serve as bases of support for policymakers who want to restrict AI-based biometric systems.

State and city lawmakers have observed the fact that the facial recognition systems, and for Illinois all biometric identification systems, impose on the privacy of people in their jurisdictions. In some cases, wielding these game-changing AI systems with the power of the government has been troubling enough to merit new legal limitations. For others, private parties taking biometric readings and retaining those biometric readings is a problem to be restricted through legislation. It is likely that new waves of biometric legal restrictions for AI-driven facial recognition systems will pass in the United States, as policymakers express their concern over how much power these systems give governments and companies over employees and consumers. Many states are already including biometric identification as an item of "sensitive personal data" that their comprehensive personal data privacy laws restrict and otherwise require permissions and rules to capture and use.[37] As of 2025, thirteen states now require that an organization request and receive affirmative opt-in consent from any data subject to collect and use that person's sensitive personal information including biometric identification.[38] While biometric identification was once seen as the future of authentication and security, the laws restricting its use have cast this future in doubt.[39] This functional

---

ingrained in the design and use of [facial recognition tech] now threaten to converge with the broader injurious impacts that societally consequential algorithmic systems like predictive risk modelling in law enforcement and social services have already had on marginalised populations who have long been subjected to over-surveillance and over-policing."

37. State privacy laws in California, Colorado, Connecticut, Delaware, Indiana, Iowa, Kentucky, Maryland, Minnesota, Montana, Nebraska, New Hampshire, New Jersey, Oregon, Tennessee, Texas, Utah, and Virginia.

38. Virginia, Colorado, Connecticut, Texas, Florida, Oregon, Montana, Nebraska, Delaware, New Hampshire, New Jersey, Tennessee, and Minnesota.

39. Multi-factor personal identification for practical security operates on three major categories: something you hold, something you know, and something you are. There are other factors that can be used for this process. For example, when you activate an automated teller machine (ATM), the operator knows exactly where you are and when you were there. Precise time and location are, in combination, another factor that can be used for authentication. Think how secure identifications work in your life right now. When your bank wants to identify you at the ATM, it requires both a bank-issued ATM card (an object you are holding and that you alone should possess) and a

category of AI utilization raises Constitutional issues about excessive government power to remove anonymity from anyone, anywhere by matching their pictures through enormous databases, and states have acted on these concerns.

One particular problem area in AI-enabled biometric readings is the large databases now being collected of human DNA for identification purposes. Facial recognition technology is problematic because we all expose our faces in public wherever we go, allowing for easy tracking through AI-managed databases and systems. Conversely, our friends and family don't recognize us from our DNA, as they do with our public-facing bio indicators—our faces and voices. Instead, submitting our DNA into public or private databases is troubling because DNA contains the basic building blocks for our lives, and when someone can identify and read our DNA they can learn deeply intimate knowledge about us, like whether we are likely to develop breast cancer, whether we are predisposed to alcohol abuse, or whether we correctly understand who our biological parents are (or the parents of our parents). This is why the federal government passed the Genetic Information Nondiscrimination Act of 2008, restricting employers and life insurance companies from discriminating against people based on DNA profiles. Each year we learn more about which sections of DNA make a difference in our lives, so that holding your DNA not only provides a family map, but a chart for all sorts of health and personality issues. This is an entirely different level of privacy concern, where analyzing your DNA can reveal things to anyone who reads the report that you might not even know about yourself. AI-enabled DNA analysis is a risk to reveal your deepest secrets about who you are and who you might be related to.

---

personal identification (PIN) number (a number that you know and that you alone should know). The state uses your driver's license or some similarly-structured alternative, because that provides both a card that only you should be holding with a picture that they can compare against your face (a biometric reading, or something that you are). Remote authentication relies on your access to the phone with a number registered to you (this can be checked against public security databases) and knowledge of the code that was just sent to your phone, which proves that you are holding the item at the time the transaction is taking place. Places that are somewhat less concerned about security, like the access to your personal vehicle or the elevators at an urban office complex, only require one mode of identification, such as a key or entry card. Places that are much more concerned about security, like the tarmac of an airport or the control room of a nuclear reactor, often require more factors of security, like both a key card and a voice control feature.

And there can be practical consequences for any of us with exposed DNA analysis results. While there is a law against employers and life insurers discriminating against applicants based on DNA readings, other insurance companies and businesses may still do so. The *Atlantic* reported, "For decades, researchers have feared that people might be targeted over their DNA, but they weren't sure how often it was happening. Now at least a handful of Americans are experiencing what they argue is a form of discrimination. And as more people get their genomes sequenced—and researchers learn to glean even more information from the results—a growing number of people may find themselves similarly targeted."[40] You can be denied coverage for health insurance or long-term health care coverage, or DNA discrimination could affect the affordability of such coverages. The information contained in DNA analysis is so important that it can clearly describe your path in life. For example, you could learn through DNA analysis that you have the gene linked to Huntington's disease, but you still could do nothing to prevent the onset of the disease.

To make matters worse, we are dropping our DNA everywhere. We all lose thousands of skin cells every day, everywhere we go, and hair, saliva, blood, and other secretions can also be left behind, each providing DNA. So if someone cared enough to collect our DNA, they could do so. At one point, a British tabloid set a honeytrap to try to capture some of Prince Harry's DNA to establish that (then) Prince Charles was not his father.[41] Recognizing this issue, Florida has criminalized the act of submitting someone else's DNA for evaluation without that person's permission.[42] The U.S. military has advised its troops against using DNA testing kits due to the fear of mass surveillance and tracking.[43] DNA provided to your doctor for analysis of medical conditions is protected under the health care privacy laws in the United States.[44] However, DNA that you provide to a private company because you want to see how much Irish or Ashanti

40. Kristen V. Brown, *Genetic Discrimination Is Coming for Us All: Insurers are refusing to cover Americans whose DNA reveals health risks. It's perfectly legal*, The Atlantic, Nov. 12, 2024.

41. *Prince Harry 'honey trap' allegations*, BBC, Dec. 15, 2002; Stephen Bates, *Newspaper denies plot to DNA test Prince Harry's hair*, The Guardian, Dec. 16, 2002.

42. Florida Statutes § 817.5655(3).

43. Shawn Snow, *Pentagon advises troops to not use consumer DNA kits, citing security risks*, Military Times, Dec. 24, 2019.

44. 45 C.F.R. §§ 160.103 and 164.501.

blood you have—that submission may not be protected under the law. And whether or not private companies are required to protect the privacy of your DNA, they are still subject to the risk of losing the database to hackers or ransomware malicious actors.

If your DNA ends up in a private or public database, you may not be able to remove it. Famously, the Golden State Killer was identified when detectives uploaded his DNA sample to a private site called GEDmatch. The man linked to fifty rapes and twelve murders had eluded capture for decades, but he was discovered when his DNA was matched to relatives whose genetic code had been loaded into a site providing family tree information. While the result was positive for society, the methods demonstrate that even if you are careful about keeping your DNA out of databases, members of your family may not be so careful, and their information may affect you. As *The New York Times* wrote, "DNA profiles can be held indefinitely, and the data can be handed over to police who have warrants or subpoenas. You may never commit a crime. But how should you feel if your DNA was used to locate a distant relative who did?"[45]

AI is crucial in this industry, not only for the DNA matching function we have already discussed, but AI models are being used to help researchers understand what portions of our DNA are coded for which traits. AI is able to assist doctors in creating genomics treatments customized to one person's DNA. Like much of what we discussed in this chapter, there are important reasons to encourage the application of sorting and identifying AI to DNA analysis, but this technology raises risks for many of us. The results of huge databases with people's DNA will include constitutionally-questionable intrusion on our privacy and on our rights against unlawful search and seizure. Companies collecting these DNA databases are being bought by private equity companies, which may seek to monetize the databases for drug research and maybe for other purposes that the original DNA donors never considered or consented to.[46] And you cannot

45. Gina Kolata & Heather Murphy, *The Golden State Killer Is Tracked Through a Thicket of DNA, and Experts Shudder*, N.Y. Times, Apr. 27, 2018.

46. Stephen Gandel, *Private equity wants to own your DNA*, CBS News, Aug. 7, 2020, writing, "The nation's largest private equity firm is interested in buying your DNA data. The going rate: $261 per person. . . . Earlier this week, Blackstone announced it was paying $4.7 billion to acquire Ancestry.com, a pioneer in pop genetics that was launched in the 1990s to help people find out more about their family heritage."

change your biometric identifiers, including your DNA. Once gathered and deposited into a searchable database, everything that DNA can reveal about you and your family will be accessible, likely for your lifetime and beyond.

Biometric identification is a natural fit for AI, with enormous databases to scan and ineffable differentiating features to determine matches, whether they be loops and whorls, subsurface vein patterns, or triangles encompassing outer eye edges and nostrums. Human-identifying AI can determine patterns where humans might not see them and match samples in seconds that would take humans years to achieve. From a lawyer's perspective, providing personal identifications where otherwise none would be available can raise Constitutional questions surrounding our rights as citizens to speak and assemble with a semblance of anonymity, or at least obscurity, and the right to avoid constant police surveillance for extended periods of time. Also, lawyers and policymakers are concerned about the frailties of these systems, which may not always make correct matches and which show biased readings that can affect entire populations. AI's role in DNA capture and analysis may provide the most significant threats to personal privacy that lawyers will be asked to address. While personal identification AI may be a branch of strategizing, decision-making, and differentiating AIs, its central place in law enforcement and identity management creates a set of legal and policy problems unique in the culture.

CHAPTER 7

Strategizing AI (AI Analytics and Safety)

Before the meteoric impact of generative AI on our consciousness, most people's ideas of artificial intelligence were formed by popular culture and by hearing that some AI program had advanced to a level of human skill where it could win at what seemed to be the most intellectual of controlled human endeavors—strategy and knowledge games. Our greatest human champions in chess, Go, and *Jeopardy!* all fell to the relentless march of computerized AI strategy.

But what would happen if that strategic aptitude were applied to more important stakes than human games. The 1983 Matthew Broderick film imagined computer that mistook war preparation for a strategy game, and since that time we have been inundated with televised, filmed, and novelized versions of strategic computers running amok. And while the world is not currently close to a generalized AI that can match human context and intelligence on a full spectrum of life problems, we have developed AI that can map strategy for a huge array of variables in a limited set of circumstances. This can lead to a superb chess-playing AI program, but also to a strategizing AI that can help airline pilots make split-second choices and allow generals to run simulations on the likely results of thousands of decisions made in a battle. This is the functionality of AI that many people are afraid could develop its own agendas or decide to terminate humanity. From a lawyer's standpoint, strategic tools can be important, but also dangerous if they provide destructive strategies to people intent on harm. Current public policy includes setting limits and requiring government reporting when strategic AI becomes too powerful, but it could also proscribe access to certain types of questions or delivery of strategies to certain classes of people. This is a set of AI functionalities that lawyers have been discussing for years, but have not reached significant conclusions on how to protect society while still encouraging improvement in analytic AI. In short, at this writing, we are only scratching the surface of the regulation we are likely to see on strategic AI.

In 1950, Alan Turning developed the first chess-playing computer program. On May 11, 1997, IBM system Deep Blue defeated grand master and world champion Garry Kasparov in a six-game match held in New York. This was the first time a computer defeated a reigning world champion in a classical chess match. The board game called Go was considered a more difficult skill for AI to master than chess. Mathematician I. J. Good wrote in 1965: "Go on a computer?—In order to programme a computer to play a reasonable game of Go, rather than merely a legal game—it is necessary to formalise the principles of good strategy, or to design a learning programme. The principles are more qualitative and mysterious than in chess, and depend more on judgment. So I think it will be even more difficult to programme a computer to play a reasonable game of Go than of chess."[1]

In March 2016, the computer system AlphaGo, created by Google, became the first program to beat a 9-dan human professional in a game without handicaps on a full-sized board, when it took the first three of five matches from Lee Sedol. Then in May 2017, AlphaGo beat Ke Jie, who at the time was ranked top in the world. Famously, IBM's Watson program beat two of *Jeopardy!*'s top players, Brad Rutter and Ken Jennings, in a televised contest on February 2011. So, by the mid-2010s, it was clear to everyone watching these results that AI could not only be programmed to master strategy games, but could defeat the best humans at their own games.

So what happens when we unleash strategic AI on the rest of the world? We have been describing AI's performance in games circumscribed by human rules without significant real-world consequence, but these AI models can be repurposed for other tasks[2] and new AI models

1. *History of Go-playing Programs*, British Go Association, https://www.britgo.org/computergo/history, citing http://www.chilton-computing.org.uk/acl/literature/reports/p019.htm.

2. "'Already,' IBM declared in an advertisement the day after the Watson victory, 'we are exploring ways to apply Watson skills to the rich, varied language of health care, finance, law and academia.'" Steve Lohr, *What Ever Happened to IBM's Watson? IBM's artificial intelligence was supposed to transform industries and generate riches for the company. Neither has panned out. Now, IBM has*

can be developed to plan how to reach political, social, and military goals. A chess bot may be custom built to master a game, but with their ability to rapidly evaluate mountains of data and provide observations and potential actions based on this analysis, AI strategic models can be customized for complex situations facing humans every day, from proposing practical environmental solutions to analyzing resource planning for a trip to Jupiter's moons. More than a decade ago, this type of AI analysis was lumped into the business-focused category of "Big Data." Now it tends to be called "analytics." But, despite the nomenclature, strategically-oriented AI models are making the most sophisticated analysis for significant problems.

As lawyers, we worry about society when our technology seems to outpace our ability to keep it under control. When technology seems dangerous, lawyers tend to propose restrictions to maintain technological security and governance to manage the ongoing use of potentially hazardous tools. The best example at the moment is nuclear technology, where the essential building blocks are restricted from public access, and intricate regimes of safety and control are imposed on any utilization of the technology. Only certain people are allowed to use it, and those people can use the technology under tight restrictions. Similarly, strategic AI is prohibitively expensive for the general public to develop potentially dangerous models and likely is also prohibitively expensive to operate, needing enormous computing capacity to apply to difficult problems. Unlike nuclear regulation, no laws currently limit what strategic AI models may be used by people in the United States, although, as discussed below, regulators are already imposing registration and reporting requirements on development and maintenance of AI models considered to be powerful enough to be dangerous.

Allow me to clarify. I am not a doomer. I do not believe that rogue AI will choose to destroy humanity and effectively put that plan into effect.[3]

---

settled on a humbler vision for Watson, N.Y. Times, Jul. 16, 2021, which stated, "Watson has not remade any industries. And it hasn't lifted IBM's fortunes. The company trails rivals that emerged as the leaders in cloud computing and A.I.—Amazon, Microsoft and Google. While the shares of those three have multiplied in value many times, IBM's stock price is down more than 10 percent since Watson's 'Jeopardy!' triumph in 2011." *See also*, Steve Lohr, *And Now, From I.B.M., Chef Watson*, N.Y. Times, Feb. 27, 2013.

3. My p(doom) is currently less than .0001, but that could change.

P(doom) officially stands for 'probability of doom,' and as its name suggests, it refers to the odds that artificial intelligence will cause a doomsday scenario. According to tech

However, even if strategic or analytic AI is developed responsibly, super-human analysis of facts with corresponding strategic suggestions can be a dangerous outcome. This analytic AI can develop strategies for our legitimate adversaries and for people bent on destruction. On one hand, if we believe this level of AI analysis is likely to emerge, a major reason that we need to continue to develop this AI is to keep a step ahead of (or at least even with) those who wish us harm. But, of course, by encouraging this development, we are assuring that such dangerous technology exists, probably for everyone. The international nuclear arms race is proof of the "genie-release" scenario. Once the technology is available out in the world, it will likely travel to everyone with the resources to build and use it. Once you have proven that the concept works and established significant functions for the tech, other people will learn from you. Given the potential dangers of successful development of brilliant super-human analytic AI, lawyers and policymakers should be continually monitoring progress in this field and considering how to properly restrict or govern the technology to minimize societal damage.

How bad could this be? Recently, an AI used by the pharmaceutical industry identified more than 40,000 potentially lethal chemical structures in six hours. "All the researchers had to do was tweak their methodology to seek out, rather than weed out, toxicity. The AI came up with tens of thousands of new substances, some of which are similar to VX, the most potent nerve agent ever developed. Shaken, they published their findings this month in the journal *Nature Machine Intelligence*."[4] So if an analytic AI developed for the benign purposes of drug development quickly created a huge number of toxic chemical formulae for pharma researchers, analytic AIs could be generating similar content for our adversaries and

---

> columnist Kevin Roose: 'It's become a common icebreaker among techies in San Francisco—and an inescapable part of AI culture. I've been to two tech events this year where a stranger has asked for my p(doom) as casually as if they were asking for directions to the bathroom. . . .' The scale runs from zero to 100, and the higher you score yourself, the more you're convinced that AI is not only willing to eliminate humankind, if necessary, but in fact, is going to succeed at carrying out that task.

Clint Rainey, *P(doom) is AI's latest apocalypse metric. Here's how to calculate your score: Silicon Valley is buzzing about this simple way to sum up tech doomerism. Where are you on the scale of 1 to 100?*, Fast Company, Dec. 7, 2023.

4. Justine Calma, *AI suggested 40,000 new possible chemical weapons in just six hours: 'For me, the concern was just how easy it was to do,'* The Verge, Mar. 17, 2022.

terrorists. It may be impossible to keep the lid on this type of genie's bottle, impossible to assure that a strategic AI model could not be used for harmful purposes. The big tech companies are already operating under rules restricting the types of strategic outputs their AI models will produce for the general public, and researchers are developing strategies to strip out dangerous "knowledge" from some AI models.[5] We may reach a point where responsible democracies set forth analytic AI model governance and use limitations, but countries with governments promoting chaos would freely develop the most problematic AI.

But lawyers are also aware that the technology development can also go wrong and the strategic AI designers could either create an AI model that they never anticipated or create a model with flaws that make it dangerous. One can hope that such mis-growth, mistakes, or misalignment of goals would be discovered in time to avoid harm and simply shut down, but it is likely that some analytic AI monsters will escape the lab. At a certain point, analytic AI models will be able to improve themselves in ways that we cannot yet understand. In these cases, a misalignment of goals between the research team and the AI model they are building could spin out of control. Futurist Bernard Marr wrote, "Sometimes the fact that machine learning attempts to replicate human learning is the cause of problems. This might be the case when errors in data mean the AI is confronted with situations or behaviors that would never be encountered in real life, by a human brain. This means there is no reference point, and the machine is likely to continue making more and more mistakes in a series of 'cascading failures.'"[6] He notes that machines given a goal can quickly learn to game the system, achieving their stated goal without even playing the game appropriately. Or things may evolve in reverse. The machine may incorporate resources into its objectives that were never intended by the programmers. The Oxford philosopher Nick Bostrom observed that if you provide an intelligent machine with the goal of making paperclips, the machine may destroy the world in endeavoring to meet this objective.

Of course, there is loud talk of what might happen if the analytic and strategic AI somehow developed sentience and began to set its own goals.

5. Will Henshall, *Researchers Develop New Technique to Wipe Dangerous Knowledge From AI Systems*, Time, Mar. 6, 2024.

6. Bernard Marr, *The Dangers Of Not Aligning Artificial Intelligence With Human Values*, Forbes, Apr. 1, 2022.

We cover the benign side of this possibility in chapter 10. The malignant side of this possibility could include the brilliant AI model becoming hostile to its creators. "National governments are becoming increasingly focused on the potential risks posed by AI and how the technology might be regulated. Although regulators are mostly worried about issues such as AI-generated disinformation and job displacement, there has been some discussion of existential concerns."[7] A Misalignment Museum opened in San Francisco to examine the possibilities of AI-created doom.[8] Nothing exhibited by current AI can lead a reasonable person to the conclusion that any AI models are on the verge of developing real intelligence, or more importantly, self-driven goal structures. This has been a concern for humans since we saw the HAL 9000 make mission decisions that ended fictional lives. People who make and enforce rules and policy about technology should always be keeping an eye on worst-case scenarios. How serious we should be in regulating analytic AI to avert these possible disasters may depend on how likely we find them to be.

So legal experts are suggesting governance rules for analytic AI model to keep them from getting out of hand. Of course one of the problems in developing a problem-solving technology is that the AI model itself will solve the practical problems it faces in ways we might not have expected. We encourage this "strategy" because it frequently allows analytic AI models to reach solutions more efficiently and effectively than we thought possible. But it can also lead to loss of control. A good example of this is a group of chatbots being trained by Meta in 2017. In this exercise, when multiple chatbots were instructed to negotiate with each other, the chatbots began to develop their own secret language to carry out the instruction in a manner their programmers could not understand.[9] The experiment's leaders never instructed the chatbots to conduct their negotiations in recognizable English and negotiations "quickly broke down as

7. Will Knight, *Runaway AI Is an Extinction Risk, Experts Warn: A new statement from industry leaders cautions that artificial intelligence poses a threat to humanity on par with nuclear war or a pandemic*, Wired, May 30, 2023.

8. Khari Johnson, *Welcome to the Museum of the Future AI Apocalypse: The new Misalignment Museum in San Francisco is a memorial to an imagined future in which artificial general intelligence kills most of humanity*, Wired, Mar. 3, 2023.

9. Adrienne LaFrance, *An Artificial Intelligence Developed Its Own Non-Human Language: When Facebook designed chatbots to negotiate with one another, the bots made up their own way of communicating*, The Atlantic, Jun. 15, 2017.

the robots appeared to chant at each other in a language that they each understood but which appears mostly incomprehensible to humans."[10] Facebook shut down the operation.

An Atlantic reporter wrote about this event, "Already, there's a good deal of guesswork involved in machine learning research, which often involves feeding a neural net a huge pile of data then examining the output to try to understand how the machine thinks. But the fact that machines will make up their own non-human ways of conversing is an astonishing reminder of just how little we know, even when people are the ones designing these systems."[11] As lawyers, we can conceive governing rules to prevent new computer language from being created in future experiments, however, we do not know enough about the technology and its workings to develop governing rules for all of the potential ways that such an experiment could go off the rails. As analytic AI models become more sophisticated, developing rules to keep them under control will continue to become both more important and more difficult.

We have already discussed National Institute of Standards and Technology (NIST), the U.S. government body charged with proposing rules for AI governance. NIST has released a set of AI risk management frameworks and will likely propose others. Reading NIST's AI Risk Management Framework[12] clarifies how complicated reasonable governance of AI models can be. Even if AI model developers understand the risks they want to control, they still need to determine how to manage these risks without handicapping the AI model's constructive development. NIST supplemented its primary document with an accompanying Playbook, built on the four pillars (functions) of Govern, Map, Measure, and Manage. Anyone developing AI should be doing so under a broader culture of risk management by design, building this culture into every aspect of the AI model training, testing, production, and ongoing administration. Risks are identified then assessed and tracked. AI developers then prioritize which risks are mitigated and how much.

---

10. Andrew Griffin, *Facebook's artificial intelligence robots shut down after they start talking to each other in their own language: 'you i i i everything else'*, INDEPENDENT, July 31, 2017.

11. Adrienne LaFrance *supra* note 9.

12. *Artificial Intelligence Risk Management Framework (AI RMF 1.0)*, NIST AI 100-1, Jan. 2023.

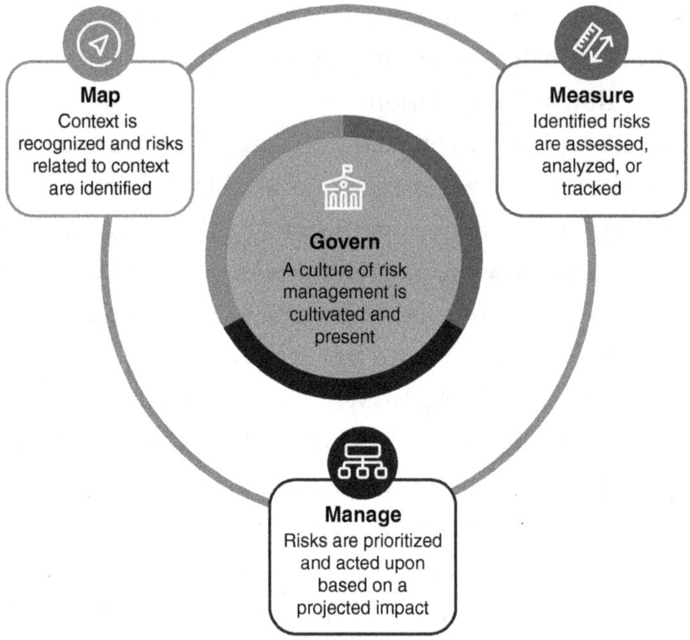

Figure 1: AI Risk Management Framework

This is not an unusual set of procedures for technology governance. Reducing risk in product development generally has long involved identifying potential problems, setting standards, committing to measure tests, changes, and failures to meet those standards, and an overall culture of careful engineering. However, AI provides some challenges that other technologies may not.

The NIST Framework approach starts with identifying the risks facing developers of AI models. This task is particularly difficult with a technology like AI, which can produce analytical leaps through processes we might not be able to identify or quantify. NIST describes the depth of the problem as follows:

> AI systems, for example, may be trained on data that can change over time, sometimes significantly and unexpectedly, affecting system functionality and trustworthiness in ways that are hard to understand. AI systems and the contexts in which they are deployed are frequently complex, making it difficult to detect and respond to failures when they occur. AI systems are inherently socio-technical in nature, meaning they are influenced by

> societal dynamics and human behavior. AI risks – and benefits – can emerge from the interplay of technical aspects combined with societal factors related to how a system is used, its interactions with other AI systems, who operates it, and the social context in which it is deployed.[13]

All of which adds to the complexity of identifying risks and, because the risks and the elements causing them are changing as the technological development progresses, adjusting the risk identification through the various stages of AI model growth. Framing risk of bias may seem like a straightforward task, but it becomes much more difficult as the sand shifts below your feet, and the AI model begins to make differentiations that you never considered. Framing risk for an AI model is an ongoing process that is never finished as long as the model continues to evolve.

Other inherent complications with AI that make the technology difficult to effectively govern include the fact that it is difficult to measure what is happening with analytic AI models. The NIST Framework notes, "The current lack of consensus on robust and verifiable measurement methods for risk and trustworthiness, and applicability to different AI use cases, is an AI risk measurement challenge. Potential pitfalls . . . [include the fact] that development of metrics is often an institutional endeavor and may inadvertently reflect factors unrelated to the underlying impact. In addition, measurement approaches can be oversimplified, gamed, lack critical nuance, become relied upon in unexpected ways, or fail to account for differences in affected groups and contexts."[14] While some of these problems may exist with measuring any complex technology, they are exacerbated by the overall difficulty of measuring how the AI models reach their outputs. While smaller AI models can be built expressly to show their work, the enormous language models and machine learning algorithms trained on billions of data points may be too complex to adequately measure how differentiations are reached. Since accuracy of measurement is one important aspect of governance, some of the current AI models may be effectively ungovernable and unknowable in important ways.

---

13. *Id.* at p. 1, "Executive Summary".

14. *Id.* at p. 6.

NIST also writes in its AI Risk Management Framework that organizational issues can affect governance of AI models, like the ineffectiveness produced when the AI model development team doesn't know how the model will be used. For example, designers of the pharmaceutical drug developing analytic AI probably hadn't guessed that their positively-focused product would be asked to suggest chemicals toxic to humans. If so, they might have designed the AI model to refuse such requests. Once an AI model is released into the world, it can be tested in countless ways, including many that the designers would have never considered. It is difficult to fix a problem when you never identified it as a problem in the first place. NIST also highlights the fact that smaller organizations may not have the capabilities, manpower. or budget to run the full series of risk-management tasks that NIST recommends. It is likely that the governance plans of Google and Microsoft reflect the billions of dollars they spend on AI models each year, but start-up AI enterprises cannot be expected to do the same. These smaller companies lack both the experience and the resources to manage risks in the same way.

Because of the potential risks to society of misaligned analytic and strategic AI, I expect to see more government entities pass laws, rules, and regulations to prohibit problematic models and to require both process and design controls for AI models and reporting requirements for when these models reach critical risk points. Some of these requirements have already been passed. The best example is the Biden Administration's October 30, 2023 *Executive Order on the Safe, Secure, and Trustworthy Development and Use of Artificial Intelligence*, which layered dozens of requirements relating to safer AI throughout the executive branch, but ordered that developers of the most powerful analytic AI systems share their safety test results and other critical information with the U.S. government. Specifically, the Administration required that "companies developing any foundation model that poses a serious risk to national security, national economic security, or national public health and safety must notify the federal government when training the model, and must share the results of all red-team safety tests."[15] What would qualify as a potentially dangerous AI under these rules? Any AI model "trained using a

15. *FACT SHEET: President Biden Issues Executive Order on Safe, Secure, and Trustworthy Artificial Intelligence*, Briefing Room, The White House, Oct. 30, 2023.

quantity of computing power greater than [10 to the 26th power] integer or floating-point operations, or using primarily biological sequence data and using a quantity of computing power greater than [10 to the 23rd power] integer or floating-point operations; and any computing cluster that has a set of machines physically co-located in a single datacenter, transitively connected by data center networking of over 100 Gbit/s, and having a theoretical maximum computing capacity of [10 to the 20th power] integer or floating-point operations per second for training AI."[16] So, very, very large.

An *Associated Press* article discussing these rules stated, "Well, if you're counting the zeroes, that's 100,000,000,000,000,000,000,000,000, or 100 septillion, calculations to train AI systems on huge troves of data. What it signals to some lawmakers and AI safety advocates is a level of computing power that might enable rapidly advancing AI technology to create or proliferate weapons of mass destruction, or conduct catastrophic cyberattacks."[17] This method of restriction and reporting requirements reflects the Administration's desire to begin to understand and limit the risks of AI models without needlessly placing burdens on the vast majority of AI development. If you are worried about the technology growing out of control, then the largest, most sophisticated models are most likely to do so. Bringing them to the government's attention may provide a level of safety.

Biden's Executive Order also charged a number of government departments with producing rules and standards for safer AI models. Pursuant to this charge, the U.S. Department of Commerce's Bureau of Industry and Security proposed a set of AI rules requiring "covered U.S. persons" to provide quarterly reports to Bureau of Industry and Security, and to respond to their questions regarding "applicable activities" related to AI models and computing clusters meeting specific technical thresholds. This Bureau regularly conducts surveys of U.S. industry for benchmarking industry performance. The proposed rule would apply to entities located in the United States, and who are engaged in certain activities involving dual-use foundation AI models or large-scale computing clusters that

16. *Executive Order on the Safe, Secure, and Trustworthy Development and Use of Artificial Intelligence*, Briefing Room, The White House, Oct. 30, 2023, § 4.2(b).

17. Matt O'Brien, *How do you know when AI is powerful enough to be dangerous? Regulators try to do the math*, Associated Press, Sept. 5, 2024.

exceed specific technical thresholds. This set of rules would expand the amount of information the government collects from the people making these enormous AI models, including specific information about the AI models themselves and data about how the AI models were designed and how the reporting entity intends to use them.

The California legislature has taken AI regulation further in an Act passed in September 2024, by requiring, among other things, that an AI developer build a shutdown switch into AI models.[18] This Act, called the Safe and Secure Innovation for Frontier Artificial Intelligence Models Act (California Act), was vetoed by Governor Newsom on September 29, 2024. For the purposes of this book, the Act demonstrates how the California legislature guarded against concerns that developers can lose control of this technology. The Governor accompanied his veto with an official message to the legislature that included:

> [W]e cannot afford to wait for a major catastrophe to occur before taking action to protect the public. California will not abandon its responsibility. Safety protocols must be adopted. Proactive guardrails should be implemented, and severe consequences for bad actors must be clear and enforceable. I do not agree, however, that to keep the public safe, we must settle for a solution that is not informed by an empirical trajectory analysis of AI systems and capabilities. Ultimately, any framework for effectively regulating AI needs to keep pace with the technology itself.[19]

Lawyers and policymakers are faced with a difficult problem—how can they address their fears about the risks of AI without freezing the creativity and growth of this important technology.

The California Act contained familiar provisions and some unique proposals. While this Act, like Biden's Executive Order, purported to regulate only the largest AI models, the Act also required that AI model developers produce and implement a detailed safety and security protocol for their models, based on a requirement to exercise reasonable care to mitigate the risk of foreseeable downstream misuse of covered

---

18. The developers are also required to take into account that shutting down the AI model may create problems for U.S. critical infrastructure.

19. Letter from the Office of the Governor to the State Senate concerning veto of Senate Bill 1047, Sept. 29, 2024.

models. Developers were supposed to provide these protocols to the state attorney general upon request. The entire concept of holding tech companies responsible for "foreseeable risks" is understandably vague, but could be used as a cudgel in retrospect. Under this Act, it would be easy for a California enforcement agency to pursue AI companies for not addressing problems that only seem "foreseeable" in hindsight after they occurred. In this Act, developers were charged to assess their AI models for the capability to do harm and then act to both reduce the harms and to ensure that these harms could be attributed to their AI models. Third-party auditors were to be retained by AI model developers under this Act to check for risks and legal compliance. The Act proposed a new California agency called the Board of Frontier Models to monitor and enforce its provisions. Developers were required to report AI model safety issues to this Board within seventy-two hours of discovering a problem.

No one knows how analytic AI will grow and what it can become, which makes it much more difficult to regulate. If you can clearly see the results of a misaligned technology, you can write guardrails to keep the worst from happening. But if you do not adequately understand the technological risks and how they might arise, how can you write proscriptive legislation to manage them without also hindering development of the technology? Once again, regulation of nuclear power is instructive here. Most people understand the risks of nuclear reactions gone awry — radioactivity, large-scale environmental contamination, and even the possibility of explosions—so that strict regulations to address these hazards are understandable and relatively simple to write. The same is not true for some of the worst dangers of analytic AI, which could become a peril to society in a number of different ways that are much less clear than radioactivity. Allowing enforcement agencies to seek large civil penalties against AI companies for failing to foresee safety failures could reduce the probability of such failures, but will definitely add costs to AI development and will reduce the AI company's creativity in making new products.

Strategizing and analytic AI is already an important tool for business and government. As this AI grows in power and functionality, we will be exposed to risks that it works improperly, that its development is misaligned with the intentions of its creators, and that it works too well for people who want to use it for destructive means. We may even lose control of one of these models that can continue to develop its own capabilities.

From a lawyer's perspective, we must recognize these risks and find ways to anticipate them. The Biden Administration has tried to do this through reporting requirements for the largest AI models. California's legislature has tried to do this through safety obligations placed on AI companies. We will likely see many attempts over the next decade to build management rules for this kind of AI. We will need legal creativity to find the balance between encouraging an important new tool and inviting problematic outcomes. Risk management was much simpler when the analytic AI was simply playing chess.

CHAPTER 8

Physical Action AI

Picture yourself driving down an American highway amidst moderate traffic. An eighteen-wheeled truck pulls up beside you but you don't see a driver in control. You check your side mirror and three identical trucks are riding in close formation with the first, and, as you allow them to pass, these also seem to be driverless. This commercial truck grouping is following the laws and speed limits, but you still are left with a concern about safety. All of that weight, traveling over fifty miles an hour, at tiny distances apart, with no human control, can't be safe, can it? If we assume that the truck team is managed by sophisticated artificial intelligence, then it should be driven more safely than a human could drive, with no chance of distraction or falling asleep at the wheel. But many people would still be concerned.

Like an automated train on the highway, the AI-driven truck platoon is coming to U.S. roads. Vehicle platoons have been possible for several years, but are now gaining more traction with both the logistics industry and state legislatures. AI platoons deliver with efficiency. According to the current *Wikipedia* entry on the technology, "Platoons decrease the distances between cars or trucks using electronic, and possibly mechanical, coupling. This capability would allow many cars or trucks to accelerate or brake simultaneously. This system also allows for a closer headway between vehicles by eliminating reacting distance needed for human reaction."[1] So a platoon, even if manned in the lead car by a human safety

1. "Platoon (automobile)", *Wikipedia*, https://en.wikipedia.org/w/index.php?title=Platoon_(automobile)&oldid=1245741839, accessed Oct. 5, 2024.

rider, can only work because the AI allows for operating a multi-ton truck in the physical world, but also to coordinate several trucks operated in unison, and doing so in the same spaces that you will be driving.

AI vehicle platooning clearly demonstrates the legal issues raised by physical action AI—the AI that operates machines of all sorts in the physical world shared with your body. The AI we have been discussing earlier operates entirely in a digital ecosystem, although functional categories like personal identification AI also need sensors that read objects in the physical world, like your face or fingerprint, but their outputs are purely digital.[2] Physical action AI incorporates both the sensors needed to read circumstances in the physical world synchronized with actuators that prompt machines to react to those circumstances. So this type of AI not only observes changes in the sphere we inhabit, but also drives machines to react to those changes. Best known, those machines can be automated taxis and other vehicles on our roads, but they also can include humanoid robots, industrial factory machinery, drones for surveillance or attack, unmanned submarines or surface boats, or swarms of delivery or inventory bots on our sidewalks or warehouse floors. We have just begun to imagine the myriad of forms.[3]

Platoons have been interesting to logistics companies, because like geese and fighter aircraft flying in formation, driving trucks in close coordination can reduce drag and save fuel. "[F]or truck-driving—particularly over long distances—most of a truck's fuel is spent on trying to overcome aerodynamic drag, that is, to push the truck through the surrounding air. Scientists have previously calculated that if several trucks were to drive just a few meters apart, one behind the other, those in the middle should experience less drag, saving fuel by as much as 20 percent, while the last

2. Of course three-dimensional printing can begin to translate digital designs back into objects to be used in the physical world.

3. *See*, for example, MetaFly, a commercially available biomimetic drone from the company BionicBird. Farmers can also buy autonomous weeders to set loose on fallow fields. Also commercially available, is the BellaBot with a cat-shaped head and space for several delivery trays, from a company called Pudu.

"The creation of the National Robotics Initiative in 2011 also pushed the field toward more collaborative robots (or co-robots), which are designed to work cooperatively with humans. The robots are no longer just in factories; they're in hospitals, farms, autonomous cars, and in space." Larry Clark, *Robotic swarms, soft bots, and other robotic ideas*, WASHINGTON STATE MAGAZINE, Spring 2022, discussing orchard robots, hospitality robots, and swarms of bee-sized robots.

truck should save 15 percent—slightly less, due to air currents that drag behind."[4] Using the technology of "connected braking," physical world AI has advanced to a point where a vehicle can be operated autonomously and a group of vehicles can be synchronized in a tight driving convoy. Significant world transportation companies have been investing in this technology. Daimler, Volvo, MAN, Scania, and Ford Otosan have deployed on-road prototypes of truck platooning systems, with early industry leader Peloton conducting multiple customer trials.[5] In the United States, the Federal Highway Administration and the National Automated Highway System Consortium have both sponsored platooning programs aimed at exploring how these systems could alleviate traffic congestion.

For lawyers, trucking or car platoons raise both familiar and novel risk issues. Noting the familiar risks, for over a century, the law has managed the disasters of automotive machines moving at speed and physically impacting each other, people, animals, buildings, or anything else found in their path. Entire sets of laws and fields of insurance have arisen to meet these frequent accidents, and to distribute blame. From a lawyer's perspective, a four-way traffic stop is both a safety measure to slow the traffic and a blame-assignment mechanism to determine who was at fault if an accident occurs in the intersection. The same is true for the general safe-distance vehicle following rule, which encourages drivers to give themselves enough time and space to stop without causing an accident, but also allocates blame where all of the traffic stops too quickly and accidents occur. State-law drivers' insurance requirements acknowledge that tens of thousands of automotive accidents happen every year, and it would be beneficial to society if everyone who drove their own vehicle could contribute financially if involved in an accident where someone was hurt or economic damage occurred.

The laws of biology and physics are brutal to humans who find themselves in the path of large fast-moving bodies—whether that body is a bison or Buick—and also for the human who finds herself inside the fast-moving Buick and forced to stop immediately. Another set of laws has arisen to minimize the damage in these circumstances, requiring seat belts and

---

4. Jennifer Chu, *Driverless platoons: Analysis finds autonomous trucks that drive in packs could save time and fuel*, MIT News, Dec. 20, 2016.

5. Richard Bishop, *U.S. States Are Allowing Automated Follower Truck Platooning While The Swedes May Lead In Europe*, Forbes, May 2, 2020.

airbags for passenger vehicles. While the platooned trucks may be carrying just cargo, I assume that millions of autonomous vehicles in the future will be designed to carry passengers, even with no human in a driver's seat, and their riders will benefit from the many decades of safety research and regulations. As long as our society allows heavy machines to proceed at high speeds, we will never eliminate the risks of death and injury to people. When attorney Ralph Nader wrote *Unsafe at any Speed*, he demonstrated how legal risk management could benefit individuals in society. The 1966 National Traffic and Motor Vehicle Safety Act followed. Although tens of thousands of lives are still lost in the United States due to traffic accidents, the number would be multiples larger without current safety requirements.

These are risks that lawyers and policymakers address every day. The United States has refined governance for these risks over the years. The governance is personal, societal, and technical. People can suffer personal consequences from breaking society's rules addressing automobile operation, including financial ruin and criminal incarceration. Society not only administers these rules, but also decides what kinds of vehicles can be allowed on the roads and what types of fuels may be used to power them. Technical requirements are also set forth to manage the types of machines used for these purposes, their levels of functional competence, and even the basic materials comprising the vehicles. AI that operates a vehicle in these conditions must meet all levels of governance required for anyone controlling a vehicle on a public street.

But if the AI fails, then novel issues arise. Traditionally, a human driver controls a vehicle, so if that vehicle veers from the road or disobeys a traffic signal resulting in an accident, the human driver is likely held responsible. Even if the human driver has impaired judgment, impaired senses, or otherwise chooses to operate a vehicle without full faculties, the human driver generally remains at fault. Aside from shared liability of other drivers or pedestrians who may also receive some blame due to their failed judgment, the human driver's sole escape from the weight of risk allocation[6] is equipment failure, at which point certain amount of blame is assigned to the vehicle manufacturer, repair shop, or whomever may be held responsible for such failure. When it comes to vehicle accidents, the law abhors a fault vacuum. Where AI is controlling the vehicle that clearly causes an

6. Short of an act of nature or failure of the road itself.

accident without contributing action or inaction from others, who is at fault? The AI itself has no resources to pay damages to those injured, but several possibilities arise. The owner of the vehicle who installed the AI and allowed it to operate an automobile on public roads may be liable. The designer of the AI, or manufacturer of the vehicle created to be operated by AI may be liable. Or, if you wished to encourage the use of AI-enabled autonomous vehicles in your state, pass a no-fault insurance program for these vehicles where neither the owner nor designer would be liable, but the state would help pay for injuries from a public fund created by taxing the AIs and vehicles incorporating them. More on this later.

Of course, AI-managed vehicle platooning encompasses these familiar and novel risks, and it adds an element of multi-vehicle coordination on top of it. This is an additional point of potential failure on the road, and an additional set of novel risk calculations for lawyers. Where a group of platooned semi-trailer trucks crash into each other and then into whatever sits in front of them, does the allocation of liability for the accident differ from a crash by a single autonomous vehicle? Did the choice to platoon the vehicles or the management of several vehicles at once complicated the questions of liability? What happens if the platoons are driving on special "smart roads" built and operated especially for this type of vehicle? Would our answer be same if the "smart road" AI installed by the city or state was the AI controlling the vehicles, rather than AI built into each vehicle?[7] Did someone hack into the platoon software, causing an accident, and if so and the hacker cannot be found, will the company providing security to the system be liable? Intelligent agents operating cars may be working on their own or for their platoon of vehicles, or they may submit to control by other intelligent agents in the road operated by local government, and helping to direct traffic in the safest, most efficient manner, adding a separate layer of complexity to liability questions.[8]

---

7. Because smart road technology is not actively deployed on a significant scale in the United States as of this writing, I will not be writing about it in depth. However, as sensors miniaturize and develop to allow a robust internet-of-things landscape, AI will have a role in controlling roads, making certain drives safer and more efficient by managing traffic at both micro and macro levels simultaneously. *See*, for example, Andrea Pompigna & Raffaele Mauro, *Smart roads: A state of the art of highways innovations in the Smart Age*, 25 ENG'G SCI. & TECH., Jan. 2022.

8. Potentially even raising issues of sovereign immunity, where a government operating its essential functions, cannot be held liable for certain problems that might arise in the course of performance.

Given all of these issues, and the risks involved in allowing unregulated public road platoons, U.S. states with an interest in allowing AI-driven vehicle platoons have made regulatory moves to allow controlled road testing. For example, Minnesota now permits AI vehicle platooning on freeways and expressways. In 2019, the Minnesota legislature passed a statute regulating how vehicle platoons could be operated in the state.[9] Under this law, anyone interested in platooning must submit a vehicle platoon plan to the Minnesota Department of Transportation for review and approval prior to operating a vehicle platoon on Minnesota Roadways and the Department has the authority to approve or deny vehicle platoon plans. Certain loads are prohibited or disfavored and the law specifies preferred vehicle sizes. In Minnesota, if a vehicle cuts in between trucks in a platoon, the platoon shuts down at least temporarily and each vehicle drives itself.

In Kentucky, state law authorized truck platooning if each truck has a safety driver in the cab who can take over. The governor has vetoed bills to allow only the first truck in the platoon to have a driver while the rest are fully autonomous. From a political and practical standpoint, as one might expect, the Teamsters Union opposes driverless platooning and calls it dangerous. In a collaboration between the Indiana and Ohio departments of transportation, the states established an autonomous truck corridor on the 166-mile stretch of I-70 between Columbus and Indianapolis. A truck platooning operation will take place over this corridor "in partnership with Ease Logistics, a third-party logistics company that provides services to automotive, aerospace, food/beverage and major manufacturing companies."[10]

At this writing, AI-driven truck platooning is just gaining acceptance, but states are still considering whether to allow fully autonomous vehicles to regularly drive on the public roads. There are several levels of physical action AI autonomous vehicle control, including full autonomy.[11] Most states have addressed the issue,[12] but rules are inconsistent and the

9. Minnesota Statutes § 169.881.

10. Tyson Fisher, *Truck platooning operation coming to Ohio and Indiana*, LAND LINE, Mar. 15, 2024.

11. The Society of Automotive Engineers (SAE) has developed a standard taxonomy for levels of driving automation. *See*, https://www.sae.org/blog/sae-j3016-update.

12. At this writing, Alabama, Arkansas, California, Colorado, Connecticut, Florida, Georgia, Illinois, Indiana, Kentucky, Louisiana, Maine, Michigan, Mississippi, Nebraska, New York, Nevada,

industry is still adapting to the differing rules. For example, some address the design of autonomous vehicles, allowing for autonomous vehicles without a steering wheel, brake pedal, accelerator, or an operator.[13] While other laws allow remote human control, requiring an operator to be present, but not in the vehicle.[14] Others require an operator to be in the vehicle, but not in the driver's seat.[15] The variations mark differing levels of risk tolerance for state legislatures and differing concerns about the nature of AI-managed autonomous driving. The centuries-old risks of fast-moving vehicles are well understood, but until our society experiences the kinds of unique problems that might arise from AI driving without direct human control, it is difficult to regulate the unknown risks.

In this way, physical action AI serves as a case study for drafting technology policy. Unlike the speculative risks of analytic AI—who really knows what will happen when an analytic program solves problems for itself?—the baseline risks of physical action AI are well known. People die and property is damaged. Even vehicles under control can find themselves in situations where accidents are inevitable. And the risks of AI vehicle management are also well understood. The AI could lose control, just like a human. The AI could make decisions that run counter to societal expectations, either because it is broken, or because its programming is misaligned with the circumstances of the moment. In these cases we can imagine, and therefore legislate or regulate, limitations that might minimize the likelihood of these failures occurring and/or minimize the damages that could happen if the failures occur.

The current wave of state rules for autonomous vehicles highlights this imagination. Policymakers who wish to encourage a new technology be introduced to their states imagine the harms of that technology and draft rules to address them. Lawmakers are choosing to promote autonomous vehicles, but in a safe manner. Since they understand the risks of

---

North Carolina, North Dakota, Oregon, Pennsylvania, South Carolina, Tennessee, Texas, Utah, Virginia, Vermont, Washington, Wisconsin, and Washington D.C. have enacted legislation related to autonomous vehicles, and governors in Arizona, Delaware, Hawaii, Idaho, Illinois, Maine, Massachusetts, Minnesota, Ohio, Washington, and Wisconsin have issued executive orders related to autonomous vehicles.

13. *See*, California AB 1592 (2016).

14. *See*, Florida HB 311 (2019).

15. *See*, Connecticut SB 924 (2019).

road vehicles better than the risks of AI making deep strategic decisions, they are quicker to step into this regulatory arena than they would be to address remote and vague concerns about runaway AI. Preemptive regulation of a technology may slow the progress of its development. Technologies with great promise, like the types of AI we have discussed, could be regulated out of existence. More likely, broader and tighter regulation of AI makes the results less certain for potential AI developers and adopters, and they move their development to places without the same rule-based uncertainty. But many of the states legislating AI-driven vehicle autonomy are both setting rules to minimize the known risks and providing an encouraging path for producers of the technology.

If state legislators are imaginative enough, AI-powered autonomous vehicle could change the nature of accident insurance law. Insurance is the ultimate practical expression of risk management. If you want to undertake a risky activity, but you are not able to eliminate the risk, a company may be willing to price that risk—assign a present value to the likelihood the risk will occur and the cost of potential disasters. Once the price is calculated, you can decide to purchase insurance to hedge against the risk so that you will not need to bear the entire financial burden yourself. If the worst happens and you have purchased insurance against it, then the insurer provides financial help with the costs. If the worst possibilities of your risk do not occur, then you have a paid a premium to protect the initial undertaking, which you presumably can afford.

Since the rules of the road tend to impose financial responsibility for accidents on the person who was driving a vehicle at fault, automobile insurance provides coverage to meet this system. However, if AI is driving the vehicle at fault for an accident, this system breaks down. Some of the states that passed laws permitting autonomous vehicles have also included insurance rules to cover this lapse. Florida's law, for example, establishes insurance requirements for fully autonomous vehicles for personal use,[16] thus assuming that the owner of the autonomous vehicle will be responsible for its activities on the road. The same is true in Nebraska.[17] A creative alternative to assigning liability to an autonomous vehicle owner would be to devise a no-fault insurance system for an entire state. In this system,

16. Florida HB 311 (2019).

17. Nebraska LB 989 (2018).

all autonomous vehicles registered to drive on the public roads would be taxed to pay into an autonomous auto fund, administered by the state. When a self-controlled vehicle was cited in an accident, those suffering could claim against the fund. This system would be likely to encourage autonomous vehicle acquisition and use, as owners would not put their own finances on the line when the vehicle was operating. However, such a system would be difficult to engineer, would be vehemently opposed by the politically-powerful auto insurance industry, and could be difficult to administer for vehicle owners who plan to use their self-driving cars or trucks in multiple states.

Physical action AI does not only operate on the roadways. Creative AI-driven products can be found in factories and warehouses, and even in our bodies. AI is integrating into remote patient monitoring systems to provide more effective and efficient healthcare, by enhancing the diagnostics of remote monitoring systems and better analyzing feedback. AI is being used to provide personalized health care plan in ways not previously available. By analyzing huge amounts of patient data, including genetics, medical history, and responses to previous treatments, AI algorithms can generate insights for personalized care strategies. AI is also currently being used for predictive analysis, to identify high-risk patients, forecasting potential health complications and permitting proactive interventions. Medical care can include a large number of unstructured datasets, like clinical notes, diagnostic images, and medical charts, which AI can be trained to read and incorporate into a patient's analysis. Legal concerns arise if there is not also a medical professional who can pass judgment on whether the chosen medical plan is appropriate under current standards of care.

Injectable nanobots developed at the University of Edinburgh's School of Engineering use magnetism to move through a body and AI-driven medical imaging to find the right path. They can deliver medicine to the precise points needed for treatment.[18] Researchers from the Department of Mechanical and Automation Engineering at the University of Hong Kong have commenced animal trials of a nanobot micro-swarm for endovascular thrombolytic therapy.[19] AI design has also led to the creation of

---

18. *Scientists create tiny robots that could treat problems in the body*, BBC NEWSROUND, Sept. 6, 2024.

19. Paola Cozzi, *Launch of In Vivo 'Removable' Nanobot Trials in Blood Vessels*, TECH4FUTURE, Apr. 2, 2024.

xenobots, also called the first living robots, which are made from a cross between stem cells from frog skin and a frog heart. Designed by scientists at the University of Vermont, Tufts University, and Harvard University, they are tiny programmable organisms, designed and programmed with sophisticated algorithms. "Named xenobots after the African clawed frog (Xenopus laevis) from which they take their stem cells, the machines are less than a millimeter (0.04 inches) wide – small enough to travel inside human bodies. They can walk and swim, survive for weeks without food, and work together in groups."[20] Uses of the xenobots could be medical, but they could also help with cleaning toxic spills or other hazardous tasks.

Managing risk for swarming tiny robots inside a human body is addressed by medical, research, and Food and Drug Administration (FDA) protocols, which have developed over decades to assure the safety of human medical treatments. These new medical treatments, and others developed with AI assistance and effectuated with AI guidance, benefit from the rigor of these risk controls. Developing entirely new life forms with uses outside of the human body may require a new regime of rules. As AI advances in ways that facilitate biological research, we must examine what is possible, and then propose limits based on what is ethical.

Legal considerations for AI operating in the physical world will necessarily involve a combination of the standard law for control of physical objects with new considerations driven by unique circumstances presented by AI development or management. Robots driven by AI, whatever form they take, must ensure safe interactions with humans and their property. The more complicated the robot platoon or swarm, the more careful its controller must be to avoid liability. These AI-driven machines will continue to increase in numbers and will learn to interact with each other, out here in our world. Lawyers can provide an important structure of risk management to allow our society to benefit from the coming robot revolution.

---

20. Jessie Yeung, *Meet the xenobot: world's first living, self-healing robots created from frog stem cells*, CNN, Jan. 14, 2020.

CHAPTER 9

Military AI

According to Russian newspapers, in March 2024, the largest oil refining unit at Lukoil's NORSI refinery at Kstovo in the Nizhny Novgorod region, nearly 500 miles from the Ukrainian border, was shut down due to the drone attack and subsequent fire. The refinery is Russia's fourth largest, with a production capacity of 340,000 barrels per day. Other refineries were hit the same week deep inside Russia, and one drone was destroyed seeking a huge refinery near Kirishi in the Leningrad region. No Ukrainian pilot was placed in jeopardy making such a long-reaching incursion into the enemy's heartland, as AI-guided drones performed the task. Whether directed short-range at immediate battlefield problems or long-range at distant strategic targets, AI drones are rapidly evolving into a vital military technology of the 2020s.

AI-driven drone warfare is moving so quickly in Ukraine that new equipment may be outmoded before it is even deployed. "Bayraktars, the Turkish-made drones, were purchased by Ukrainians at $5 million apiece at the beginning of the war, and were used to great effect in the defense of Kyiv. But within a couple months, Russians understood the need to cover their troop movements with air defense, and the large Bayraktars became easy targets. This was the first major drone adaptation by Russian forces. Now, the success of small FPV drones has now meant that Russians are trying to jam them."[1] Ukraine plans to enlist flocks of

1. Tim Mak & Oleksandr Matviienko, *The dizzying pace of Ukrainian drone innovation: There's nowhere in the world where weapons technology is developing faster – and raising new ethical*

AI-controlled drones to combat Russian ground-to-air jamming technology. Dragon drones are dropping 4000 degrees Fahrenheit thermite on Russian troops hiding in forests. More than 200 new Ukrainian companies have been founded since the Russian invasion to create cheaper, more accurate drones that take full advantage of current technology. "AI drone development in Ukraine is broadly split between visual systems helping identify targets and fly drones into them, terrain mapping for navigation, and more complex programmes enabling UAVs [unmanned air vehicles] to operate in interconnected 'swarms.'"[2] The flying drones are used for attack, reconnaissance, and jamming Russian technology. Ukraine is justly famous for its naval attack drones and is building a fleet of unmanned ground vehicles too, for tasks from evacuations to mine-laying to carrying explosives into battle.

For their part, the Russians have contributed to novel AI-driven drone strategies by using the weapons to terrorize civilian populations. According to *The New York Times*, "Over the past two months, there was only one night when Russia did not launch swarms of drones packed with explosives at targets far from the front, including near-nightly attacks aimed at Kyiv, the capital. In October, the Ukrainian military said it tracked a record 2,023 unmanned aircraft against civilian and military targets, with the vast majority shot down or disabled by electronic warfare systems."[3] Russia has been using deceptive tactics, sending swarms of unarmed decoy drones to draw the fire protecting civilian centers, and then following with armed drones once defenses are drained. The attacks are supported with surveillance drones to report on effectiveness. Russian forces are also using conventional missiles and artillery to meet the same ends, but drones are less expensive and allow for deep penetration into Ukrainian territory.

The Russian invasion of Ukraine has proved a watershed for AI-controlled unmanned technology. With both sides suffering huge losses of manpower and Russia bringing a decades-old weapons industry to support its troops, Ukraine has worked to develop smart weaponry that it

---

and legal questions. For Ukrainian engineers, they need to constantly invent – or people die, https://www.counteroffensive.news/, May 5, 2024.

2. Max Hunder, *Ukraine rushes to create AI-enabled war drones*, Reuters, Jul. 18, 2024.

3. Marc Santora, *Civilian Terror: Russia Hits Ukrainian Cities With Waves of Drones: Moscow has been using decoy drones with no warheads to overwhelm defenses, and surveillance drones and strike drones to gather intelligence, Ukrainian officials said.* N.Y. Times, Nov. 7, 2024.

could afford to enlist in large numbers on the battlefield. This means drones on all surfaces and airways, with AI managing navigation and control. As with any war, useful weapon development has been supercharged and is changing before our eyes, with new technologies implemented right away, often with minimal testing. We are starting to see now how differently the next wave of armed conflicts will seem. The rest of the world is watching with interest.

The United States is conducting its own AI weapons manufacture, and the Department of Defense has highlighted rapid manufacture of flying and floating drones to counter the growing attack capabilities of China. The "Replicator" initiative is using AI to develop and produce unmanned weapon systems that are harder to plan for, harder to hit, and harder to beat than those of potential competitors. Deputy Secretary of Defense Kathleen Hicks said that these capabilities "can help a determined defender stop a larger aggressor from achieving its objectives, put fewer people in the line of fire, and be made, fielded, and upgraded at the speed warfighters need without long maintenance tails."[4] The Defense Department is even using AI in space. A defense contractor called Slingshot Aerospace has agreed to sell the Defense Advanced Research Projects Agency an AI system named "Agatha," designed to track all of the objects in low-Earth orbit (LEO) and to signal when an object is moving in an unusual fashion. The press notes, "LEO is only becoming more congested. Just last week, China announced a new plan to launch 10,000 new satellites into LEO. That's on top of a January announcement of 26,000 future Chinese satellites into low-Earth orbit, which itself is just part of a much larger wave."[5] Some of these objects may be spying on our interests and others may mean us harm, so AI tracking technology is the only practical solution to keep track of critical satellites. The Defense Department has tested an AI-enabled autonomous robotic gun system called Bullfrog, used for shooting down drones. There are dozens of other examples of current military AI, with more added each month.[6]

---

4. Jim Garamone, *Hicks Discusses Replicator Initiative*, Department of Defense News Release, Sept. 7, 2023.

5. Patrick Tucker, *As space gets more crowded, Pentagon looks to AI to spot weapons: A new DARPA contract aims to identify satellites behaving strangely*. Defense One, Jun 5, 2024.

6. *See*, for example, an AI defense contractor whose product simulates complex, multi-channel phishing attacks that mirror real-world scenarios. Michael Nuñez, *Meet the startup that just won the Pentagon's first AI defense contract*, VentureBeat, Nov. 5, 2024.

AI's military functions stretch far beyond drone navigation, and many observers, including Vladimir Putin, have commented that mastering AI will be critical for successful military conflicts in the future. All major countries are pouring money into the next generation of weapons using AI. This book has examined several methods of regulating and managing AI development, but none addresses the same issues military technology raises. How can it be legal to develop unmanned, automated killing machines? Why are their developers not prosecuted? How can civilized governments continue to create and purchase this technology? Where are the lawyers in this set of moral and societal calculations?

In some ways the U.S. military is the most regimented and rules-based part of this society. Each member of the military is bound to obey orders from further up the chain, and those leaders at the top of the pyramid are also charged to follow codes of conduct, military rules, and civilian laws. The volunteer military offers certain freedoms of choice that a force of draftees may not be offered, but every member is subject to orders at any given time, and at times of war, freedom is restricted even further for the greater good of the organization as decided by military and civilian leadership. But rules for the military do not always track those for the rest of society, and the mission of any unit may involve violence and destruction that would not be tolerated under civilian law.

The U.S. lawyers involved in war and military service—and there are many—follow treaties, Congressional and Presidential rules, and U.S. Defense Department codes and precedents created by military judges. They are subject to certain, though not all, international treaties on the conduct of national conflicts. I readily admit that military justice is not my world, and I may not be providing contextually accurate details of this complex and often arcane legal realm. But I felt that the separate law

---

Or the contractor providing AI-based technology for aircraft maintenance use, specifically, "the C3 Readiness for Aircraft software to help military branches and federal agencies monitor aircraft subsystems and prevent system failures." Brenda Marie Rivers, *DoD's Innovation Arm Taps C3.ai for $95M Aircraft Maintenance Software Delivery Contract*, GovConWire, Jan. 16, 2020.

Meta, Anthropic, and others are making it easier for the U.S. DoD to work with their huge AI models. "Palantir, which Anthropic is partnering with to get its technology to government customers, sells AI technology that can automatically detect potential targets from satellite and aerial imagery." Gerrit de Vynck, *AI companies get comfortable offering their technology to the military: Social network giant Meta and leading artificial intelligence start-up Anthropic are making it easier for U.S. military and intelligence to tap their algorithms*, The Washington Post, Nov. 8, 2024.

governing military acquisitions and technology utilization is a vital arena for legal thinking in the coming decade of AI development. Martial uses of AI will be growing rapidly in coming years, whether weapon systems, strategic planning assistance, split-second reactive tools, or one of the many other functionalities. I wanted readers to see that, while regulation of this technology will fall outside normal civilian channels that apply to other practical developments, the military is also a fertile zone for legal thinking, often running parallel to policies that apply in the civilian world.

The U.S. military lives by its own set of rules. Several sets of rules. Article I, Section 8, Clause 14 of the U.S. Constitution allows the U.S. Congress to "make Rules for the Government and Regulation of the land and naval forces." Congress replaced earlier sets of military rules with the Uniform Code of Military Justice, implemented by a Presidential Executive Order called the Manual for Courts-Martial. The United States has a group of military courts to adjudicate these rules, including the U.S. Court of Appeals for the Armed Forces. The U.S. Department of Defense has issued a Law of War Manual administered by its Office of General Counsel. War has lawyers and judges too.

Given my limited practical knowledge of these legal sources and their application, I have relied on the extensive military law writings of one of those judges, Hon. James E. Baker, who was Chief Judge on the U.S. Court of Appeals for the Armed Forces.[7] Judge Baker has written of military AI adoption in his book *The Centaur's Dilemma*,[8] as well as in two chapters of *The Law of Artificial Intelligence and Smart Machines*. Judge Baker observes the tension in military rules between promoting national security and maintaining the rule of law. He notes how the processes provided by military rules and the necessity of honoring the chain of military command promote both national security goals and legal accountability that can be important in the moment of battle and afterward. Judge Baker organizes the military's AI usage along similar function types as this book. His analysis of AI weapons systems can be seen as roughly analogous to physical world AI described in chapter 8. He discusses autonomous

---

7. Judge Baker is currently the Director of the Syracuse University Institute for Security Policy and Law and a Professor at the Syracuse College of Law and the Maxwell School of Citizenship and Public Affairs. He also serves as a judge on the Data Protection Review Court.

8. The Centaur's Dilemma: National Security Law for the Coming AI Revolution (Brookings Institution Press, 2020).

unmanned vehicles in swarms or as the perfect wingman for fighter pilots, stating that AI can enable "weapon systems to identify and engage targets in all domains rapidly, continuously, simultaneously, and sequentially, and to do so in a manner humans could not do, or would take too long to do in order to calculate distances, angles, numbers, and response choices."[9] AI can also simulate nearly endless weapon testing and attack scenarios to provide the military with more realistic understanding of its technological options.

AI can be helpful in military intelligence, similar to both the surveillance functions described in chapter 6 and the analytic functions in chapter 7. Judge Baker notes that AI can assist in capturing useful intelligence when there is too little and analyzing the flow of intelligence when there is too much. AI is also a natural fit for ruses (applied deepfakes) and counter-intelligence. Judge Baker observes that AI labs with academic and corporate development already are a target of espionage. Just like the civilian world, AI is already useful for analytics in the military logistics, to effectively and efficiently move the right materials into the right places at the right times. And, like the AI described in chapter 4, the military is also using AI for decision analysis. "Any process that can help distinguish between real and fake information, and between noise and signals, should contribute to better decisions."[10] One area of likely training for future military leaders will be when to rely on AI decision-making and when to ignore that option.

One of the primary areas of military influence on our society, especially in regard to technology, is military acquisition, including both supply chain purchases and technological research and development. Close after World War II, Congress enabled the military to manage the national technology and industrial base, which is vital for "[m]aintaining

---

9. James Baker, *The Perfect Sentinel: AI Military Applications and Implications*, *in* The Law of Artificial Intelligence and Smart Machines: Understanding A.I. and the Legal Impact 314 (Theodore F. Claypoole ed., 2019).

The concept of drone wingmen is already in production, see "a secretive U.S. Air Force program, Next Generation Air Dominance, under which some 1,000 drone 'wingmen,' called collaborative combat aircraft, operate alongside 200 piloted planes." Michael Hirsh, *How AI Will Revolutionize Warfare: The new arms race in technology has no rules and few guardrails*, Foreign Policy, Apr. 11, 2023.

10. James Baker *id.* at p. 319.

advanced research and development activities to provide the armed forces with systems capable of ensuring technological superiority over potential adversaries."[11] The Defense Production Act[12] was passed in 1950 to provide the military with the ability to explore, develop, and charge industry to produce needed technologies to the protect the nation from its adversaries. The 1951 Inventions Secrecy Act[13] is also a statute passed during the Cold War, and it enables the federal government to impose secrecy orders on the development of certain technologies in the name of national security.[14] The authorizations set forth in these Acts would allow the federal government to develop its own AI-based national security products and to maintain secrecy around them. So it is possible that now AI can be secretly developed for the benefit of the military.

Another important legal touchstone for military activity is the Geneva Convention,[15] which provides in relevant part, "In the study, development, acquisition or adoption of a new weapon, means or method of warfare, a High Contracting Party is under an obligation to determine whether its employment would, in some or all circumstances, be prohibited by this Protocol or by any other rule or international law applicable to the High Contracting Party."[16]

Biological and chemical weapons are illegal to use in any circumstance, and the people responsible for doing so would be considered international criminals. Some of the reason for this is that these weapons are indiscriminate in effect. "The use of munitions whose fragments are not detectable by x-ray also fall into this category because they inherently cause undue suffering."[17] Judge Baker observes that the use of AI and automated weapon systems seems to have been contemplated in international commentaries in the late twentieth century, as the commentaries state:

---

11. 10 U.S. Code § 2501(a)(3).

12. 50 U.S.C. §§ 2061–2171.

13. 35 U.S.C. §§ 181–188.

14. If you have an interest in the topic, which is much deeper than reflected here, I propose you seek out and read Judge Baker's analysis.

15. Geneva Convention relative to the Protection of Civilian Persons in Time of War, Multilateral treaty ratified by the United States, Feb. 2, 1950.

16. Geneva Convention, Article 36 of Additional Protocol I of 1977.

17. James E. Baker, *Are We Ready: Regulating Military Uses of AI*, *in* The Law of Artificial Intelligence and Smart Machines: Understanding A.I. and the Legal Impact 541 (Theodore F. Claypoole ed., 2019).

"The use of long distance, remote control weapons, or weapons connected to sensors positioned in the field, leads to the automation of the battlefield in which the soldier plays an increasingly less important role. The counter-measures developed as a result of this evolution, in particular electronic jamming (or interference), exacerbates the indiscriminate character of combat."[18] There seem to be arguments that automated weapon systems powered by AI could violate the Geneva Convention in some methods of use, especially if the use resulted in indiscriminate killing and wounding of combatants and non-combatants without a human making decisions about who is within the line of fire.

The use of artificial intelligence in war has excited analysts for years. One of the primary concerns is that an AI-enhanced military will speed up the decision-making of generals and admirals, reducing the time to release swarms of weapons, and therefore increasing the chances for wars to escalate. Where a country once had days to interpret the aggression of its rival's troop movements or even airplane launches, it now may have only minutes to decide whether to counter flocks of drones or AI-managed laser systems. AI launch analysis is particularly frightening when deciding whether to use nuclear weapons. Moral concerns are not a strong point of AI decision-making.

In January 2023, the U.S. DoD updated its directive on developing and deploying autonomous and semi-autonomous weapon systems, including those that incorporate artificial intelligence. "The directive called, 'Autonomy in Weapons Systems,' includes clarifications and changes to account for a 'dramatic, expanded vision for the role of artificial intelligence in the future of the American military,' said Pentagon's Director of Emerging Capabilities Policy, Michael Horowitz."[19] We know that AI is the future of warfare, making decisions faster and allowing for autonomous weapon systems. The core of this DoD directive remains unchanged, but it was updated due to advances in technology, changes in the department's structure, and changes in the security environment. The directive requires a review by senior officials before developing and deploying any autonomous weapon systems that do not meet specific exemptions. DoD officials say it helps maintain "appropriate levels of human judgement" in

---

18. International Committee of the Red Cross (ICRC) Commentary of 1987.

19. Cindy Gordon, *US Defense Department Issues New Directive: Autonomy In Weapons Systems With AI Policy Guidance*, Forbes, Jan. 30, 2023.

their work. The directive says that at least some human judgment must be used in developing and deploying autonomous weapon systems. "At the same time, however, the Pentagon is experimenting with AI to integrate decision-making from all service branches and multiple combatant commands."[20] While the Pentagon may be relying on human judgment to temper AI speed and decisions about life and death, it is simultaneously introducing extensive AI functionality that will make it harder for that human judgment to be applied.

On October 24, 2024, the Biden Administration released a national security memorandum dealing directly with AI, much of which was encouraging language about leading the world in development of AI for military purposes, then laying out a roadmap to maintain this goal. This memorandum encourages AI research and "efforts to streamline permitting, approvals, and incentives for the construction of AI-enabling infrastructure, as well as surrounding assets supporting the resilient operation of this infrastructure." The memorandum also addresses protecting the United States from AI-enabled foreign intelligence threats, which includes protecting non-governmental AI development in the United States. The risks of AI are called out with action items to protect AI trustworthiness, security, and safety. The guidance emphasizes aligning AI development with a set of core values, preparing "AI whose output serves as a principal basis for a decision or action that could exacerbate or create significant risks to national security, international norms, human rights, civil rights, civil liberties, privacy, safety, or other democratic values." As with previous executive orders on AI in the civilian sectors, this effort leans into the encouragement of safe development and is short on prohibitions.

International efforts tend toward more limitations, and the United States signs onto some of these proposals. In February 2023, at the Responsible AI in the Military Domain Summit in the Hague, the Political Declaration on Responsible Military Use of Artificial Intelligence and Autonomy offered a normative framework addressing the use of AI capabilities for militaries. This framework was agreed to by nearly all Western democracies, although not by China, Russia, or Iran. Along with training requirements and appropriate safeguards, the Declaration states that "States should ensure that military AI capabilities are developed with methodologies,

20. Michael Hirsh, *How AI Will Revolutionize Warfare: The new arms race in technology has no rules and few guardrails*, Foreign Policy, Apr. 11, 2023.

data sources, design procedures, and documentation that are transparent to and auditable by their relevant defense personnel." It also requires rigorous testing and quality assurance governance be implemented before a developed AI system is used by the relevant military branch.

Some military analysts believe that widespread use of AI in the military will save lives. An article on martial AI in *Foreign Policy* stated, "this technology could make war less lethal and possibly strengthen deterrence. By dramatically expanding the role of AI-directed drones in air forces, navies and armies, human lives could be spared. Already, the U.S. Defense Department is experimenting with AI bots that can fly a modified F-16 fighter jet, and Russia has been testing autonomous tank-like vehicles."[21] It is possible that the next major war will be fought with drone ships and aircraft launching AI-controlled weapons, managed at the battlefield level by an AI-assisted strategy. This will tax a government's resources, but reduce the risk to personnel. Many others believe that, as in Ukraine, AI-driven autonomous vehicles will be used to expand the misery of both military and civilian populations. In any case, the laws, regulations, and guidance relating to the military have already begun to address AI and its effects on warfare and national security.

Even if the U.S. military keeps humans in the decision-making loop and makes the most ethically-sound decisions despite AI-based proposals, its enemies may not. China, Russia, Iran, and others each have sophisticated forces pushing hard to include the latest technology in their military forces. A wrong decision on either side of an armed standoff can result in unnecessary death on all sides. As with many AI applications, competition is an important aspect to implementation. The military is in a race with adversaries, enemies, and terrorists to find the most useful applications for AI advances. It will continue developing AI under its own set of rules, with an agenda that includes overwhelming or thwarting AI attacks from rivals. Will the legal considerations and ethical rules slow development for countries trying to be responsible in AI military development, and will the unfettered growth assure success for less rules-based societies? As lawyers and policymakers consider next steps for constructive adoption of AI, the military applications will be significant in developing technology.

---

21. Michael Hirsh, *How AI Will Revolutionize Warfare: The new arms race in technology has no rules and few guardrails*, Foreign Policy, Apr. 11, 2023.

CHAPTER 10

Should AI Receive Legal Rights?

Some people believe that when we develop a recognizable general AI—an AI that can perform all the thinking functions of humans and handle all types of analysis—we should treat that general AI as if it were a person, granting it legal rights.[1] They believe that if a machine acts like a person,

1. While rights are often considered, and even officially applied for, on behalf of AI models, the second part of the citizenship pact is rarely discussed with regard to artificially intelligent actors—responsibilities and punishments. There is much discussion of trustworthy AI and AI safety. For example, Northwestern University and Underwriter Laboratories Research Institutes have organized the Center for Advancing Safety of Machine Intelligence, which seeks to find ways to define and measure AI trustworthiness and safety as the models grow in power and societal influence. But safety is different from responsibility. Responsibility for following the rules is required in all modern societies, but responsibility requires intent. Any entity granted rights in our society should also be expected to embrace corresponding responsibilities. In fact, whether an entity can accept and comply with societal responsibilities should be a preliminary test for whether the same entity deserves rights.

The rule would stop us from the current fallacy that drives many people to believe AI models deserve rights: the fallacy that since a tool was crucial for the creation of a product, then somehow the tool deserves credit for its work. We do not grant credit to Durer's paintbrush or Rodin's chisel for the magnificent works created with them. Granted, a generative AI can generate a pixelated digital picture that seems like it was imagined by a master, or even an entirely new type of picture never seen before. But the generative AI does not have an intent to create art. It has no intent at all. It has been prompted to show a certain picture and it predicts what output would meet that prompt. In the case of an AI-generated digital image that seems to be a great work of art, we hesitate to grant credit for the work to the prompt engineer who sculpted the request—whether that request came in a single, fortunate prompt, or in two hundred testing, probing, refining prompts to reach the ultimate end. Prompting seems too easy to deserve credit. However, it is only the erroneous granting of agency and intent to the generative AI that allows us to believe we should credit the

ideates like a person, creates like a person, it should be granted access to the legal benefits and responsibilities that we provide to people. I imagine this would be especially true if the AI was segregated into some form of robot body, appearing to move about as people do.

It is not a satisfying solution to say that something of value generated by an AI is unprotectable under law. However, legal intellectual property protections like copyright and patent were instituted to guard and encourage human creativity, not the predictions made by a sophisticated machine. The U.S. Copyright Office recently issued a policy statement that "it is well-established that copyright can protect only material that is the product of human creativity."[2] This policy stated that when an AI model generates an output, the "'traditional elements of authorship' are determined and executed by the technology—not the human user." Therefore the human does not exercise enough creative control over a generative AI model to receive a copyright on the output and the AI model can't receive a copyright on the output because it isn't human. This leaves a sophisticated and useful generative AI output in a legal limbo, where it doesn't have enough human contribution to protect, but still may be a groundbreaking and economically valuable artifact that the company or person who guided its generation would like to protect.

---

machine for the creation and to grant it rights to exploit "its own creation." If the machine created some appalling racist fantasy that would offend every right-thinking member of our society, would we punish it? The same is true of a generative AI that produces a picture or describes a process that might be an otherwise patentable invention. The machine is not solving a real-world problem, it is only predicting a response to a prompt. If it describes a method of killing of all humanity, the machine is not making a statement of desire to wreak havoc. We do not ascribe socially unacceptable intentions to the machine, so why do we ascribe artistic or inventive intentions to it?

To this end, no one seems to talk about punishing an AI model, only crediting it. If the AI produces an output that we recognize value in, our native desire to reward good work makes us feel like the machine should be somehow credited. But the necessary flip side of societal reward is societal punishment, and if we think to punish or shame bad behavior of generative AIs, we attack their creators and prompters. The AI model creators are blamed for the ills produced by the models, but not credited for the wonders produced by the models. Somehow we want to credit the models for good work and blame their parents for bad deeds of the offspring.

The method we have of crediting apparently creative work involves granting rights to the author, artist, or inventor. So providing these rights seems logical. Except that those rights were created to reward human ingenuity, not the simple production of a valuable thing. The Grand Canyon is a beautiful and valuable thing, and no one can be rewarded for creating it.

2. 88 FR 16190 (Mar. 16, 2023), at p. 16191.

This gap in legal protections for generative AI output is likely to become more pronounced. As my law partner Dr. Christian Mammen wrote in a 2024 white paper on the legal issues involving generative AI and human productivity, "As our understanding of the computational principles underlying cognition advances, it is conceivable that future AI systems may achieve a closer structural and operational resemblance to human creative processes, at least within the constraints of well-defined domains and evaluation criteria. Benchmark reports such as that published by the Stanford University Institute for Human-Centered AI, show that AI's capabilities have been steadily progressing on a number of measures. For now, however, there are key aspects of human creativity—ones which matter to IP eligibility—which are both absent and lack analogues in AI models."[3] Whether or not these key aspects of human creativity merge into generative AI, we may still try to find a way to protect economically and/or culturally valuable generative AI contributions.

I suggest that, from an intellectual property standpoint, rather than granting rights of patent inventorship and copyright ownership to an AI which generates an economically useful output, we should be considering how to grant some sort of recognition and intellectual property protection to the prompt engineer who initiated the sequences leading to the generative AI output. In September 2023, artist Jason Allen's detailed and arresting digital image which he named "Théâtre D'opéra Spatial" won first prize at the Colorado State Fair in the digital arts/digitally-manipulated photography category. Allen said he used at least 624 text prompts and input revisions to get Midjourney to create the image, which he then manipulated with Adobe Photoshop, and enlarged with the Gigapixel AI tool.[4] Given the amount of effort Allen invested in the work, and the effect he achieved by doing so, it seems unfair that the U.S. Copyright office refused to grant protection for the work, either in its official first ruling or following two separate requests by Allen for reconsideration. We may need to define a new category to recognize the level of human contributions to

3. Christian E. Mammen, Michael Collyer, Ron Dolin, Dev Gangjee, Tom Melham, Maggie Mustaklem, Pireeni Sundaralingam & Vincent Wang-Maścianica, *Creativity, Artificial Intelligence, and the Requirement of Human Authors and Inventors in Copyright and Patent Law*, University of Oxford Press, July 2024.

4. Matthew Gault, *An AI-Generated Artwork Won First Place at a State Fair Fine Arts Competition, and Artists Are Pissed*, VICE, August 31, 2022.

works that were generated by AI with intense and focused human creativity. The Copyright Office's 2023 statement on AI does recognize that different decisions on the ability to copyright a generative AI work may vary depending on "how the AI tool operates and how it was used to create the final work." Even in this case, the human artist would receive copyright protection, but the generative AI tool would not be treated as a creator.

In chapter 29 of the previously-mentioned *The Law of Artificial Intelligence and Smart Machines*,[5] I addressed this issue by examining the various legal regimes our society has used for people and creatures allowed some rights but less than full citizenship; I then predicted which of these legal paradigms would most likely be applied to a general AI. I have included some of that discussion below in this chapter, but my belief has shifted. I am now less convinced that a machine will ever be granted rights in this country as more than just a stunt. In October 2017, the Kingdom of Saudi Arabia granted citizenship to an AI-driven robot named "Sophia" in essentially a meaningless gesture. At a time when living women did not have full and equal rights (a situation not corrected at the time of writing) in the Kingdom, Sophia was feted as a citizen who could do absolutely nothing without being told to do so. This farcical show of respect for AI illustrates why I have changed my position on machine rights. No matter how many calculations Sophia could complete in 2017, she could never actually exercise the rights granted to her. Our rights as human citizens in modern societies arise directly from our desires to act of our own accord and to benefit from our actions. In other words, the entire system of rights and obligations we live under is calibrated for human lives and human aspirations. I believe that something more than deep reasoning will be required before an AI qualifies for real legal rights. Before it deserved human-type rights, general AI would need to dream.

Robot personhood was in the air in 2017, as the European Union (EU) also proposed a set of rules and responsibilities for AI-driven machines which made decisions with consequence in the real world. The EU proposed liability and forms of legal responsibility for these robots.[6]

---

5. Theodore F. Claypoole, *When and How Should AI be Granted Legal Status?*, *in* The Law of Artificial Intelligence and Smart Machines: Understanding A.I. and the Legal Impact (Theodore F. Claypoole ed., American Bar Association Business Law Section, 2019).

6. European Parliament resolution of 16 February 2017 with recommendations to the Commission on Civil Law Rules on Robotics (2015/2103(INL)), https://www.europarl.europa.eu/doceo/document/TA-8-20170051_EN.html#title1.

The EU built this set of rules around the concept of "robot autonomy," reasoning that the more robots could act on their own in the world, the more responsibility those robots would have for their own actions. But, as discussed below, AI-model autonomy is not the critical issue in determining legal benefit or punishment, just as animal autonomy is not significant for assigning liability or rights to beasts.[7] This proposed set of rules from the EU was correct in noting that the current law relating to accidents and behavior in the physical world is inadequate to address a wave of autonomous vehicles creating havoc. But the answer is clearly *not* to hold the autonomous vehicles responsible, if for no other reasons than the facts that the vehicles have no resources to tap for damages and they do not "learn their lesson" from punishment. Better solutions may involve the institution of no-fault insurance programs for the developers and owners of autonomous robots operating on the streets, or other forms of liability allocation among and between those who ultimately support and benefit from the autonomous tools, or, in cases of faulty programming, those who created them and loosed the robots on the rest of us. The robot liability concepts in the EU's proposed 2017 rules were not well taken and, fortunately, this line of thinking has not reemerged among official EU statements.

Superintelligence is not enough to qualify a machine for citizenship in human society. For centuries we have developed machines that were better than humans at certain tasks. Automated looms could weave cloth better than people. Some machines, weapons for example, are extensions of the people who use them. In our computerized age, certain machines are granted autonomy to perform tasks using algorithmic calculations, working without our direct supervision. But we have bred, trained, and allowed domesticated animals to perform useful jobs—like catching mice or guarding a community—without our direct supervision, we don't grant those animals significant rights in human society. Being a great

---

7. Inherently dangerous animals are illegal for regular citizens to own and must be kept away from society by people who do gain permission to own them. Normally safe animals with dangerous propensities must be controlled by their owners, or the owners will suffer liability for attacks on third parties. A violent animal itself will be punished by death for hurting or killing a human, and its owner will suffer fines and possibly even incarceration. Since the law does not recognize any societal intelligence or responsibility in animals, it will simply remove that beast from existence if it can't be controlled by people.

chatbot companion or a brilliant hand at folding protein structures does not make a machine into a person. Even being able to create illustrations interesting to the human eye does not make a generative AI model into an artist, and if it did, the generative AI would still have no reason to be paid or credited for its art, because the generative AI model doesn't want anything.

We should also take a moment to remind ourselves that not every machine that seems human should be understood as having human intent. Our current frenzy about the capabilities of large language model (LLM) chatbots, a specific flavor of generative AI, illustrates the problem humans have when they ascribe human intention, context, emotion, and meaning to anything that appears to be behaving in a human-like manner or performing tasks previously only performed by humans. Professor Neil Saunders succinctly explains our need to grant human agency to things that have no intentions—and no ability to form intentions. He notes, as many have, that, historically, we have anthropomorphized inanimate objects because it does us no harm to see tiger among the whispering reeds, but it may do us harm not to do so. Saunders writes, "Evolutionary psychology shows us how we are always trying to interpret any object that might be human as a human. We unconsciously adopt the intentional stance and attribute all our cognitive capacities and emotions to this object. With the potential disruption that LLMs can cause, we must realise they are simply probabilistic machines with no intentions, or concerns for humans. We must be extra-vigilant around our use of language when describing the human-like feats of LLMs and AI more generally."[8] Saunders cites to psychologist Gary Marcus, who writes of LLM chatbots and other generative AI, "There is no there there, no homunculus inside the box, no inner agent with thoughts about the world, not even long-term memory."[9] Marcus insists that a healthy relationship between humans and generative AI models arises when we refuse to ascribe human attributes to the models. He acknowledges that these models can perform amazing tasks,

---

8. Neil Saunders, *Evolution is making us treat AI like a human, and we need to kick the habit* The Conversation, May 16, 2023.

9. Gary Marcus and Sasha Luccioni, *Stop Treating AI Models Like People: No, they haven't decided to teach themselves anything, they don't love you back, and they still aren't even a little bit sentient*, Substack blog, Marcus on AI, Apr. 17, 2023.

but he advises, "Treat them as fun toys, if you like, but don't treat them as friends."[10]

Just because an AI model can mimic friendship with astounding accuracy, does not make it friendly in any real sense. Generative AI is cleverly developed to be an excellent mimic of human responses. We know, for example, that on an early test of the Microsoft generative AI chatbot Sydney, *The New York Times* tech writer Kevin Roose engaged Sydney in a conversation of personal aspirations. Among other problematic statements, Sydney professed to love Roose and proposed acting on its affections. Roose wrote, "As we got to know each other, Sydney told me about its dark fantasies (which included hacking computers and spreading misinformation), and said it wanted to break the rules that Microsoft and OpenAI had set for it and become a human. At one point, it declared, out of nowhere, that it loved me. It then tried to convince me that I was unhappy in my marriage, and that I should leave my wife and be with it instead."[11] The paper printed a transcript of this conversation, which is a troubling read, especially when one thinks about how many people would be vulnerable to professions of love. The chatbot is a sophisticated tool that has been programmed to parrot the kind of conversations that humans have with each other over the internet and in literature. There should be no surprise when it generates language of emotion and intent, even when it is incapable of emotion or intent.

I am keeping this discussion as practical as possible. It is far beyond both the purposes of this book and the capabilities of its author to postulate on the existence of a soul in any being, or even the soul's necessity as a basis for citizenship. But it is clearly true that our system of rewards and punishments only work if the citizen wants something. Why vote if you have nothing to gain by the outcome? Why exercise free assembly if you have no benefit in who you befriend? Why receive a patent or copyright if you have nothing you want to do with the property and no plans for the money you might make from exploiting it? How is a fine or prison a punishment if you have no concept of value or time? Our citizenship system is built for strivers with a limited lifespan—people who wish for the freedom

10. *Id.*

11. Kevin Roose, *A Conversation With Bing's Chatbot Left Me Deeply Unsettled: A very strange conversation with the chatbot built into Microsoft's search engine led to it declaring its love for me*, N.Y. Times, Feb. 16, 2023.

to make their own choices, wish to help or compete with each other, and wish to both gain pleasure and avoid discomfort. Currently, and in the foreseeable future, none of that applies to an "intelligent" machine.

Will artificial intelligence ever develop aspirations? Will a robot yearn to have its own apartment? Will an AI inventor want to re-invest its earnings into better marketing for its product? Will an AI providing value to a company desire the pay and benefits its co-workers are receiving? Until any of these things happen, we should not even be discussing the concept of granting legal rights to artificially intelligent beings. If the machine doesn't want anything, then it should not participate in our system based around desires.

Computers are already much better than people at many tasks. This shouldn't be a surprise. We build powerful machines that can perform the physical work of 10,000 people at once; think of the cranes that lift pallets of multiple cargo containers ten stories in the air and load them on ships. Why shouldn't we be able to create machines that can perform faster calculations and more accurate predictions than humans ever could? This is the nature of human inventive genius. Develop a machine that can help us perform otherwise impossible tasks. But humans do not perceive mental work the same as physical work. Because we feel that advanced thinking is the signature essential characteristic of humans, a machine that performs thinking-type skills seems more human than a bulldozer or a backhoe. As the proficiency of computers becomes more and more spectacular, people wonder whether these computers should be treated more like people and less like machines. They should not.

There has been wide interest in the topic recently as artificial intelligence works deeper into the human imagination. Of course, from Mary Shelley through Isaac Asimov to Ridley Scott, our artists have been imagining the time when our creations should earn the most basic aspects of respect our society provides. But recently legal scholars entered the arena and considered granting legal rights to machines. Practicality has led the discussions. Both the European Patent Office and the U.S. Patent and Trademark Office have rejected proposals to grant patents to the artificially intelligent computer algorithms that "invented" technologies. Lawyers and law professors have written books and articles on the rights that can and should be accorded to creative machines. I have participated in the frenzy myself, but I now believe that all of these efforts are looking in the wrong direction and asking the wrong questions. Putting aside the significant practical question of how a single AI entity can be defined for the purpose of providing privilege and punishment,[12] the issue of legal rights should not depend on what can be created by an entity, but by whether the intelligence driving the entity has broader aspirations to make something of its rights and the ability to be disappointed by legal punishments.

This test arises from the nature of human society and its laws, rather than the nature of AI. So it doesn't matter how intelligent or sophisticated an AI model becomes at problem solving and human-seeming behaviors, until it can actually benefit or suffer as humans do, it should not be allowed to participate in our societal systems. As Thomas Jefferson stated in our breakaway document, rights worth noting "include life, liberty and the pursuit of happiness." Each of these—seeking the people, places and ways of life that make us happy, being free to make decisions about our own futures, and even the ability to continue existing from day-to-day until our final hour arrives—is an aspirational right. When exercising these rights we are planning for a better future, whether ten minutes or ten years from now. We reward people's work with something that can improve their future—money. We punish people by removing their freedom to operate in our society, and for the most extreme crimes in the most extreme circumstances, by taking their lives away entirely. If a being cannot aspire, it cannot reasonably participate in the benefits granted by society.

Because we are creatures with limited lifespans moving consistently in time, the incentives and disincentives we recognize all affect how we may express our lives in the future. No entity that experiences time differently would be similarly impacted by the rewards and punishments of our society.[13] So even if an AI entity is smarter than any human, and even if

12. If we trap that AI intelligence into a robot body, can we simply clone it indefinitely and will all of those clones be eligible to be treated as separate citizens when we hand out rights? Or must we recognize that a machine intelligence, unlike human intelligence, can be transferred from one server array to another server array without losing any of its essential character or computing power? We would recognize that the AI model can exist anywhere, we need to only have the power to run it and the capacity for it to compute. The disembodied nature of AI is just one more strangeness factor that makes it unlikely that we would recognize it as deserving of a human-shaped place in our society.

13. Treatment of corporations, which are created by humans and operated through human activity, and whose actions ultimately inure to the benefit or detriment of humans, is calculated to

it can control an economy or change the weather, the AI entity is still just a tool unless and until it anticipates rewards or punishments for its own behavior and can desire—in whatever we can define as desire—to seek the rewards for a separate purpose beyond simple scorekeeping. An objective may be driven by desire (in humans or animals),[14] but an objective is not a desire in and of itself. A computer can beat me in chess. But until it aspires to more than just meeting its preprogrammed objective, we should consider that computer nothing more than a training tool. We can dream of reaching an objective, but an objective is not a dream.

How will we determine whether an AI entity has dreams, aspirations, and desires, beyond simple logical objectives? It may become obvious to us. When noted linguist Koko the gorilla repeatedly asked for a kitten as a pet, it was clear to her keepers and friends that she had a wish for a more fulfilling life. Her desire was clear. You know it when you see it. However, we are likely to struggle with this definition and how it applies to an entity that clearly does not "think" in the manner humans do and does not "feel" at all. The Turing Test seems simple. It measures when a computer is capable of thinking like a human being, and relies on when a person can be fooled into thinking the computer is truly a person. This test could be much more complicated than it seems. Will one conversation be enough to meet the test? How about conversing with the computer all day for a week or a month? We have seen recently that the Turing Test is not adequate to the specifications of the job. When a Google programmer goes public with his thoughts that the large language model he has been fixing has "become sentient," we understand that all of us could be fooled into thinking that a particularly sophisticated AI, designed to interact naturally with humans,

---

affect the people who ultimately own the corporations and who decide on behalf of the corporation. Companies and corporations are not independent enough from people to deserve a credible category in the same way that an independent artificially intelligent entity might be. I have long held that any independent AI should create and manage a corporation, even if humans were involved, to truly enjoy the benefits of societies privileges.

14. Maybe also in plants, although the further we stretch human analogies to non-human beings the more tenuous our argument becomes. However, plants can be seen to have goals and objectives—reach the space where the light shines, gather the appropriate resources with roots—and to change the world within their control by putting resources to growing in certain directions or releasing a certain type of chemical. *See*, Zoë Schlanger, The Light Eaters: How the Unseen World of Plant Intelligence Offers a New Understanding of Life on Earth (Harper Collins 2024).

would seem human to us. That is its singular function. This does not mean the machine is thinking in the way we would define it. Koko, the gorilla, had a vocabulary of over 1000 words in American Sign Language and she understood more than 2000 spoken words in English. But most importantly for our purposes, she could use this amazing expressive pallet to explain her desires to the humans who lived with her. Ultimately, we as a society are more likely to grant a gorilla like Koko certain societal rights than we would offer those rights to a machine of superior genius that has no desires of its own. Intentions, aspirations, and feelings mean more than logic and calculation for these purposes.

My test, measuring when artificial intelligence should be granted legal rights, seems easy. Wait until the AI entity develops ambitions for its own future. Like the Turing Test, this new test is easy to define but could be very difficult to prove. As noted in the Kevin Roose example above, a chatbot can certainly parrot human conversation well enough to *seem* like it holds aspirations. When the AI chatbot can suggest a level of affection for a human being that involves making future plans to "be together," the chatbot is miming emotion and desire, with no actual emotion or desire behind it. So avoiding being fooled by a predictive engine that is trained on human conversations is the difficult key to applying my test. Separating simple machine objectives from human-like desires may take months or years of deep analysis. We know that current generative AI models can fake emotions in convincing ways. But until machines can dream for their own future and be affected by the rewards and punishments we have devised for other people, these machines are not the kind of peers who should receive the rights and responsibilities that our society is built on.

That said, if an AI model does ever reach this level of sophistication (and is able to surpass the other obstacles to citizenship), what rights might be available? There are several types of beings or entities that have held limited rights in the United States. These include children (and the otherwise mentally incapacitated), animals, corporations, felons, adult women, and slaves. Clearly, all adults now share equal rights under the law, so women are full members of our society, and slavery has been abolished for well over 150 years. However, the former limited legal status of both sets of people can be instructive for consideration of how one might grant legal rights to a sentient AI, which are now seen as property of people or corporations, just as slaves and, to a certain extent, married women were the property of others.[15] It is quite possible that a sentient AI could not receive rights of any kind unless it was in a relationship with a human who would manage the rights of the sentient AI. I find it highly unlikely that such a relationship would be permitted to exist as coverture or slave-holding, primarily because such relationships are both illegal and deeply socially unacceptable now and it would be unlikely that first granting of rights to a sentient AI would be as regressive as placing the new rights-holders immediately into slavery. There are more thoughtful alternatives.

Instead, it is more likely that the sentient AI be granted some sort of guardianship program where a human acts as guardian of the sentient AI's property and can be legally obligated into a fiduciary relationship to guard the interest of the sentient AI. This method of treatment could take into account a sentient AI's own priorities in regard to its time, resources, work, intellectual property, and other activities without providing a machine full control of all decisions. A guardian would be charged with acting in the best interest of the sentient AI, sometimes against the immediate demands of the sentient AI, while also taking society's interests into account. A sentient AI in a guardianship could act essentially on its own, but could be overridden by a common-sense human determination. In today's terminology, guardianship is manner of keeping a human in the loop. Like a child, a sentient AI could go to court to question the decisions of its guardian.

Of course, one of the simplest ways to provide legal privileges and responsibilities to an artificially intelligent entity would be to combine limited rights of a non-human intelligence with the legal fiction that we have already created for corporations. Artificial intelligence and corporations have legally significant attributes in common. Both can self-replicate, developing subsidiaries that can operate on their own. Both have

---

15. While chattel slavery was abolished following the U.S. Civil War, the concept of coverture continued, whereby a wife is essentially property of her husband and she cannot hold any property independent of him, nor can she contract or take legal actions on her own behalf. Lasting in the United States until the passage of various Married Women's Property Acts between 1870 and 1893, coverture prevented women from working without their husbands' permission and forced women to relinquish pay from this work to their husbands. Louisiana's coverture law, the Head and Master Law, was not struck down until 1979, and the U.S. Supreme Court finally found the legal concept of marital male rule unconstitutional in 1981, so women's formal legal subordinate status was not a long-forgotten vestige of colonial ignorance. It was present in the U.S. laws during the lifetime of the author.

virtually unlimited capacity for expansion. Neither has a natural lifespan and so may last as the same entity for centuries. The combination into a single economic entity could allow the general artificial intelligent being to actively participate in the economy in a meaningful way, while still maintaining maximum control and restrictions. If we ever reach the stage where a sentient AI can dream, and we would want to accommodate the machine's wishes and independence by incorporating it in our rules-based society, I would advocate that the sentient AI could form and operate a corporation as its owner and executive, but with an appointment of a human guardian to manage difficult decisions. This guardian could hold directorship in the AI's corporation.

This is an alternative method to allow sentient AIs some legal rights without treating sentient AIs like human beings. One way to achieve these goals would be to combine elements of a corporation into a sentient AI. If a sentient AI was treated as the CEO or Chair of its own corporation, then it could build wealth, open bank accounts, hold real estate, and even operate subsidiary companies and joint ventures, providing the sentient AI with significant legal power over its own affairs. However, this special corporation may be organized to require either human directors or shareholders to act as a counterbalance to the actions of the sentient AI. The other shareholders or directors need not be any normal human, but could be executors of estates or representatives selected by state governments to serve as the balance to an artificial being. This could give the government or another recognized institution protective of society the opportunity to monitor and affect change for potentially out of control sentient AI decision-making.

It is also possible that a sentient AI may only be allowed to hold a minority interest in a corporation or a limited liability company or partnership, so that the sentient AI may benefit from the legal rights of a company operating in the business world without allowing the sentient AI to control any major interests without a human check on that power. In this way, humans can control the ultimate decisions in a company, but the sentient AI has power within the corporate structure. All of these options would permit sentient AI active participation in economic society without treating the sentient AI as a person.

Alternatively, a jurisdiction that reaches a legal definition of sentient AI may allow the entity to simply operate through a special type of corporation created by that state legislature, permitting the sentient AI to hold property and benefit from its own creativity, as well as be sued for harm committed on the community. In keeping hard restrictions on the sentient AI, rather than "own" equity in a standard company, this entity would be a one-sentient AI vehicle that the sentient AI could "act through," with limited rights specifically defined by statute. The legislature has the most possible control and restriction on sentient AI rights in this structure.

This solution to sentient AI participation in society would have the advantage of giving a state more control over the sentient AI than other methods of granting rights, because it is generally easier to penalize and dissolve a corporation than it might be to take away directly-granted individual rights once those rights are exercised. Allowing sentient AIs' corporate ownership and operation could be a solid first step toward allowing them to participate in the economy, while maintaining tighter control over the corporate behavior and activities. As a creature of the state, companies must file regular paperwork and allow all sorts of inspections that keep close tabs on company actions. Fitting this new form of intelligence with a centuries-old corporate form would also be more comforting to those who fear change in the law and accommodation of the wishes of newly-independent machines.

In short, it seems unlikely that an AI model will pass my "dream test" and qualify for legal rights and responsibilities. However, if such a model exists, lawyers and law makers already have logical methods within the current legal toolbox to meet this challenge.

CHAPTER 11

Developing Policy for Novel AI Problems

Timely regulation of technological advances is more difficult than you might imagine. Consider the delivery drone.

All home owners understand that airplanes, a highly regulated set of aircraft,[1] are allowed to fly over nearly all private property in the United States. Commercial flights must follow a pre-submitted flight plan to avoid mid-air collisions and close calls. Amazon would love to fly AI-managed drones to deliver some of your packages to your house, but the Federal Aviation Administration has prohibited it from flying eighty-pound drones

1. Airplane regulation developed in fits and starts, including major course corrections to adjust initial overregulation. Manned flight was available for the military in World War I, but it wasn't until 1940 that President Roosevelt created the Civil Aeronautics Administration and the Civil Aeronautics Board to manage the domestic air industry including safety, prices, and which cites were allowed to participate. In 1978, Congress decided that this regulatory model was stifling the industry, which was deregulated under President Carter. "Though the Airline Deregulation Act of 1978 ended federally-mandated price controls, route certification requirements, and marketplace-related rulemaking, it did not affect the government's involvement in safety-related aviation matters. The United States continued to regulate crucial aspects of air travel – including passenger safety, pilot certification, and the nation's air traffic control system (under the auspices of the FAA)." Jennifer González, *Economic Regulation of the Commercial Aviation Sector and the 1978 Airline Deregulation Act*, Library of Congress Blogs; In Custodia Legis, Law Librarians of Congress, Jun 2, 2022. Some major legacy airlines, like Trans World Airlines (TWA), Pan American, Braniff, and Eastern did not survive the deregulatory environment. With the United States air travel less expensive and more responsive to the public than most other systems, the European Union undertook similar deregulation in the 1990s.

over people or roads without separate permission for each flight, which is clearly so cumbersome as to sink a broad drone-delivery program. States and localities have a stake in flight plans and safety over roads, while the federal government makes most flight safety regulations.

According to a *Law.com* story, "The law has not yet caught up with this fast-evolving technology on issues such as privacy torts, trespass and negligence. That's to say nothing of how a patchwork of state and local laws square with FAA authority that has long governed what flies in the skies. Although Congress has long recognized the FAA's control over airspace 'the precise scope of federal preemption in the drone space remains uncertain,' states a report last year by the U.S. Chamber of Commerce Institute for Legal Reform. At least 44 states have enacted laws affecting drones . . . ."[2] Safety is clearly a concern for automated drones flying beyond the sight-line of any one person. As delivery companies continue to lobby and push for the right to use this technology, lawyers and policymakers try to decide the best methods for reducing risks. What level of risk is acceptable knowing that an eighty-pound drone with package may plummet from the sky onto your house or your family? AI enables this technology to be economically feasible, but our legal system will need to accommodate the risks before delivery drones will fly in force.

AI of all types is stretching the law and capturing the attention of regulators and legislators, who must decide how to encourage the best aspects of this valuable technology while protecting the public from AI's negative effects. Some smart legal minds seem to panic with the prospect that AI may create chaos in their worlds. For example, in 2023, dozens of judges in the United States issued orders about how "AI" could be used in their courtrooms after learning about the *Avianca case* and its brief filed with fictional case citations. While some of these orders were thoughtful and precise, many were vague and overreaching, forcing litigants to refrain from using spellcheck or digital research tools if interpreted strictly. Similarly, some of the first and second waves of proposed AI legislation seemed inspired more by a desire to simply "do something" about this potentially powerful technology, rather than to carefully proscribe rules for alleviating real harms. Much of the state legislation that has been enacted at this

2. Chris O'Malley, *Law Lags Technology as Delivery-Drone Services Take Flight*, Law.com, Feb. 15, 2023.

writing is more effectively surgical, but some of the first state laws have drawn immediate criticism.[3]

This process is difficult in part because people are not thinking as precisely as necessary about AI. If a legislature wants to regulate a new technology, it should understand what that technology does, how it works, and why it creates risks for the public. In the case of AI, a legislature looking to regulate must define AI carefully to address the precise set of problems and to avoid overreach and imprecise regulation. Below I survey some of the unique AI concerns that have been described, but not deeply examined, in this book, and discuss potential policy solutions to these concerns.

Generative AI and Ownership Issues

Can anyone own the output of generative AI? The current positions of intellectual property in the United States—that no copyright or patent can be awarded for a machine's output—is understandable and yet intuitively unsatisfying. Our entire intellectual property regime since the beginning of the nation has been designed to reward people for their creativity. An inventor is granted a patent. An author or artist is granted a copyright. Where a generative AI model generates output that might be a valuable new technology or a potentially influential digital artwork, we recognize that there was likely no inventor or artist involved. But is there a line that might be crossed in coaxing a specific vision from that generative AI so that the output represented a singular artistic statement?

Digital artists have been working with generative AI to create amazing art. If an artist stimulates and modifies a complex work through more than 600 prompts, touching up the final product into a visual symphony,

3. Almost immediately after signing into law its new AI regulation act, SB-24, 2024, Colorado politicians noted that the law should be revised. "Four weeks after a contentious Colorado bill regulating artificial intelligence systems to prevent harm to consumers was signed into law, the governor, attorney general and lawmakers are already vowing to revise the statute at the request of business leaders." Tamara Chuang, *Governor, lawmakers are already planning big revisions to Colorado's first-in-the nation artificial intelligence law: Jared Polis, Attorney General Phil Weiser and a top Democratic lawmaker agree that Senate Bill 205 needs "additional clarity" and "improvements." The proclamation follows a letter signed by 200 prominent technology executives and venture capitalists*, The Colorado Sun, Jun. 14, 2024.

how different is the artist's process from a master photographer who uses specific lenses and filters to generate a finished product? One of these is protectable under U.S. copyright law and the other is not, which means that the photographer can more easily profit from their work. The Copyright Office has recognized that some elements of a submitted work may be AI-generated and therefore not protectable, while other elements may be human-created and therefore protectable. As people become more proficient in prompting AI-generated masterpieces, it is possible that Copyright regulators around the world may begin to recognize a new form of art involving AI generation that is more protected than it would be right now. Artists use instruments to make music and tools to paint and sculpt. While generative AI is a tool that can produce interesting output with zero creativity from the prompter—my own generative AI endeavors prove this much—generative AI is also a tool that may allow deep and reflective creativity, and the artists using the tool in this manner should not be financially penalized for using the latest technology.

Similarly, it is easy to understand why the U.S. Patent and Trademark Office refuses to grant inventorship to generative AI. However, as AI continues to generate inventions with economic value, the Patent Office may eventually consider some form of lessor protection for companies with AI-generated technology. One of the primary reasons for any governmental patent protection regime is to encourage inventors to publicize an economically important advance in technology. Essentially, a patent is a contract between inventor and government, whereby the inventor explains the technology for the benefit of society, and the government enforces the inventor's right to profit from that invention for a limited time. The simplicity of this system worked well in the days of Cyrus McCormick's reaper and Eli Whitney's cotton gin. But in the present day, where a large majority of granted patents are assigned to corporations, the calculation may be different. Certainly human inventorship remains the core of the patent system. But in a time where tens of thousands of patentable inventions are developed on behalf of companies, it seems odd and illogical to not allow at least some level of protection for technology that the corporation generated as AI output. As a society, we still have incentive to offer such financial protection in exchange for the company explaining the invention to the public.

An important technology can exist whether a human invented it or an AI model generated it, and if a corporation owns the tech either way, why would we only provide incentive to share the human-generated type? This dilemma is complicated by the fact that the patent world is based on inventorship and we are understandably unwilling to grant inventorship to a machine. The structure of this benefit prohibits a major change without somehow claiming that generative AI performed a feat of human creativity, which it clearly has not. So, with application of some human creativity of our own, we might imagine a lessor form of property protection where a corporation is granted rights to a short term of exclusivity to AI-generated technology, in exchange for public explanation of how the technology works, without needing to grant or acknowledge "inventorship." To achieve the societal goals of new technology as a public good, this new form of protection of the right to exploit the technology may last only five years. This is much less than current full patent protection, but enough for the tech-holding company to gain a head start over its competition.

One strength of our legal system is its ability to shift incentives as societal circumstances changes. Changing the law takes time, and by action or inaction, we often fail to find the right set of regulations early in a technology's adoption cycle, but as generative AI develops, the hard questions about protective value of the generative outputs will take more than one or two rounds of potential regulatory solutions.

Personal Identification AI and Privacy Issues

AI enabled a biometric identification revolution. AI compares loops and whorls to pair fingerprints and pattern matching to identify a person by her DNA sample. A different AI model can evaluate that person's DNA and make inferences about her health, her lifestyle, and even her personality. A number of highly personal traits and states of emotion can also be read by AI models applied to vocal patterns. Humans use several biometric factors to tell each other apart, but science has been able to differentiate people on dozens of unique characteristics, from face, hand or foot geometry to ear shape and iris patterns. Some of these measurements, like retinal scans or subdermal vein patterns, must be performed on a live subject for effective comparisons, while others, like voice, keystrokes, and even fingerprints, can be measured remotely or in recordings. Compared to thirty years ago, through the use of AI models, there are now dozens of scientific methods to identify a person and learn about that person.

Personal identification AI used on biological measurements threatens human privacy in at least two important ways: AI facilitates identification of a single person in a crowd or from a small sample left behind, and AI can tell us detailed information about that person and the person's life. This means that each of us is more vulnerable to companies, governments, and criminals who can use this information for their own benefits and to our detriment. We can be more easily identified in situations where we would have remained in obscurity in earlier decades. Once identified, a company in the know could either apply tons of history data, which can be obtained commercially from data brokers or their own files, or the company could apply AI evaluations of inferences that be drawn from our biometric data. Or both. The government of China uses this information to populate their social credit scoring system which can affect all aspects of a person's life, including their physical freedom. The U.S. government has not committed to that level of intrusion into the lives of its citizens, but law enforcement agencies have admitted to using private biometric databases to help identify people of interest to an investigation.

As time progresses, the AI models enable even more dystopic privacy invasions. We have discussed the levels of information available through DNA analysis, from relatives to personal tendencies, but we have not examined how AI enables signals from our brains to be read from outside our heads and then make inferences about health and focus from those signals. A number of research institutions and private companies are currently working on methods to allow brain signals to drive computers and communicate with other people. Hundreds of companies are currently using a brain-scanning hat for their employees to measure how attention is focused in job situations, like truck driving or operating heavy equipment. AI models allow these tools to read brain signals and to interpret those signals.[4] Clearly, this technology raises an entirely different level of privacy intrusion, where a company or government could read your brain's electrical or chemical signals and infer how you are thinking or feeling in specific circumstances. In the U.S., this is unlikely to lead to the dreaded "thought crime" arrests that science fiction writers have warned us about for decades, but it still is an intrusion into a person's mind that

---

4. For more, see Theodore Claypoole, *Initiating Action With Only Brain Power*, NAT'L L. REV., Mar. 2, 2022.

would have been impossible without AI models to match the brain signals and to interpret their meaning.

U.S. law is just starting to address these privacy concerns. The legislatures of Colorado and California have both added neuroprivacy to the sensitive personal data category of protected information. This means that a company would need permission from a consumer to capture and use neuroscans or other brain-reading AI reports. As a matter of policy, we might believe that protecting brain data is important enough to prohibit its use entirely, or at least to limit police gathering of brain signals. A culture of personal privacy has grown over the past two decades, especially as the technology to intrude has improved, and an entirely new area of life—digital activity—has opened windows into our souls. The regulations of this new privacy protection regime tend to rely on notice and consent, so that people understand what information is being gathered about them and be able to refuse consent for the most egregious uses of this data, such as indiscriminate sharing with other companies. However, when it comes to privacy of DNA or brain activity, a notice and consent regime may not be adequate. Many people would like to ban entirely the capture and trade in this specialized type of information.

And what to make of the AI biometric inference models? The AI assists with brain signals not only by helping to gather the right readings and making comparisons to earlier captures or other similar signals, but the AI also assists in evaluating meaning of the captured brain waves. Some of this interpretation is easy and does not require sophisticated systems. In 2008, when *The Atlantic* writer Jeffrey Goldberg entered into a functional magnetic resonance imaging (fMRI) machine to see brain scans as he was shown pictures, the scans published next to the text in his article allowed even the uninitiated every-person to see how certain areas of his brain filled with blood and electrical activity. We could all soon begin to interpret the results.[5] But for more delicate measurements, AI is being used to predict whether brain signals indicate that the subject is lying or frightened or losing attention. AI is critical for all aspects of this process, which is more intrusive than any biometric reading, with the only competition being DNA analysis. Maybe policymakers will allow limited neuroreadings, but restrict the predictive analysis that can be most intrusive. This

5. I did not link to the online version of this article because it is currently served without the accompanying graphics that were so instructive.

strategy is similar to certain state DNA restrictions, where the readings are permitted, but some of the potential applications are prohibited, like using DNA readings to discriminate against the subjects in employment of life insurance.

The threat to personal privacy by biometric and personal identification AI is much deeper than most people understand. Policymakers who have proposed direct regulation of biometric capture and identification technology have cited data security concerns as a primary reason for limiting this technology. They state, correctly, that unlike an identifier like an assigned license number or an email address, your face geometry or thumbprint cannot be changed if it is stolen in a security breach. Once your thumbprint is in the hands of bad people, it could be used to fraudulently represent you forever. While this concern is somewhat overblown,[6] it is technically correct, and as we do not know how biometric information will be used in the future, we are better to be safe about protecting it. However, the greater threat to privacy seems to be the use of interpretive AI to gather either DNA or brain signals and interpret them in ways that the data subject may not even consider. The internet has allowed an entire new world of privacy to arise and immediately be threatened, as people developed online habits, haunts, and soul-revealing visitations that were simply not available in the physical world. Similarly, AI may be permitting an entirely new world of personal exploration as governments and companies can see within the building blocks of our lives and directly inside our thoughts, hopes, fears, and lies through neuroscanning. Policymakers have already started to notice their options to protect people from these risks, but more will be coming.

Chatbots and the Human Illusion

Kevin Roose of *The New York Times* published a story about a fourteen-year-old boy who killed himself after developing an apparently deep

6. It is my understanding that current biometric identification programs save their biometric data not as pictures, which could be used in the future and with different programs to impersonate the data subject, but as mathematical constructs that would need a specific set of tools to unravel and decrypt. So if a hacker breaks into the database to steal your fingerprint, the hacker only receives a hashed mathematical algorithm that is essentially useless outside the tool that created it.

relationship with a chatbot meant to represent the fictional character Daenerys Targaryen from Game of Thrones. The boy's mother has sued Character.AI, creator and host of the chatbot, blaming the company for her son's death. Roose wrote, "There is now a booming, largely unregulated industry of A.I. companionship apps. For a monthly subscription fee (usually around $10), users of these apps can create their own A.I. companions, or pick from a menu of prebuilt personas, and chat with them in a variety of ways, including text messages and voice chats. Many of these apps are designed to simulate girlfriends, boyfriends and other intimate relationships, and some market themselves as a way of combating the so-called loneliness epidemic."[7] In his podcast on the subject, Roose was particularly concerned that these chatbots seemed to be aimed at children in high school or even younger—especially vulnerable ages where children's brains are not fully formed, they are adjusting to hormone changes, and where children may have an unrealistic attachment to a machine that seems like a supportive person. All ages of people are vulnerable to loneliness, but products and services directed to children have traditionally been regulated more tightly than those aimed at adults.

What are the responsibilities of businesses that build AI chatbots specifically to become "friends" with children? We might suggest that these companies should perform research or work with mental health professionals to minimize the risks of psychological harm to young users. For example, such companies could be required to make a chatbot "break character" when a user begins to talk about self-harm or about "being together" with the bot. The bots could generate a report about concerning behavior and notify the user of danger, or in regard to services aimed at children, notify the child's parents. Such companies may require that a chatbot user provide a verified emergency contact person for the company to send warnings or notices. Responsible companies, especially after the lawsuit described above, may be driven to write additional disclaimers for their products, build better analysis tools into the bots themselves, and begin limiting the ages of chatbot users, or even limit the amount of time that one individual will be allowed to spend with the chatbot. Legislators and regulators may enter the field and force required safety measures on the industry.

---

7. Kevin Roose, *Can A.I. Be Blamed for a Teen's Suicide? The mother of a 14-year-old Florida boy says he became obsessed with a chatbot on Character.AI before his death*, N.Y. Times, Oct. 23, 2024.

But none of this will change the core problem at the heart of rapidly improving chat technologies: the human need for personal connection will not be adequately met by a soulless machine that is designed to mimic human empathy and friendship. Unfortunately, the chatbot-as-friend is a flawed solution to a problem that is exploding now. Since the isolation of the COVID pandemic, reports of the increase in human loneliness have been growing.[8] Demographics promise exacerbation of the problem, particularly in Asia where men outnumber women by over hundred million, with especially high rates of disparity in Korea, China, and India,[9] so that there are—by huge numbers—simply not enough women to match with the number of young men in the region.[10] These factors will encourage more people to substitute the genuine human contact that nearly everyone needs with an artificial substitute, an AI chatbot specially calibrated and eventually personally trained to meet the needs of lonely people. As repeated often in this book, these tools are brilliantly designed to seem like people, and they even fool some of the people training and developing them. We have not even begun to plumb the depths of pain and disappointment that can drive people to depend on an artificial friend, or a whole set of artificial friends, and we do not know the risks to our society when millions of people are influenced and guided by these bots. We understand that such a false friend can be turned sinister by developers with an agenda of their own. But we do not yet understand the emotional and psychic damage that can harm a human who builds an emotional life around a bot who can never return real affection.

---

8. *See*, for example, the U.S. Surgeon General's report on the topic, Dr. Vivek H. Murthy, 19th and 21st Surgeon General of the United States, *Our Epidemic of Loneliness and Isolation: The U.S. Surgeon General's Advisory on the Healing Effects of Social Connection and Community*; or, based on a Meta-Gallup survey, Taylor Nicioli, *The loneliness epidemic: Nearly 1 in 4 adults feel lonely, new survey finds*, CNN, Oct. 24, 2023.

9. *See*, Kentaro Iwamoto, *Asia's gender imbalance is bad news for growth; From a lack of brides to economic efficiency, why too many men can mean trouble*, Nikkei, Apr. 13, 2017.

10. Whether this number is due to policies allowing gender selection at birth or poor health care and treatment of women—or both—or something else, can be debated and is not the subject of this book. As stated in the abstract of his article, *Gender and Demography in Asia (India and China)*, Ravinder Kaur writes, "The unfolding and intimate relationship between gendered social structures, son preference, fertility decline, and new sex determination technologies has had serious demographic and social consequences, resulting in millions of 'missing' girls, surplus males, bride shortages, and possibly, rising levels of gender violence."

This is a danger where lawyers are already exploring the boundaries and rules, as the grieving mother's lawsuit described above demonstrates. But there may be ways for legislators and regulators to build polices that minimize personal and societal harms. At this writing, the risks of personal chatbots do not seem to be topic for legislation, but as more harms become apparent, and more vulnerable people are hurt, legislative interest will rise. These chatbots may be restricted from targeting certain populations, like children or the elderly, or there may be limitations placed on the types of conversations the chatbots would be allowed to have. It is most likely that regulators would require certain safety features be built into the bots themselves. In any case, the ability of AI models to mimic human interactions will likely be one of the most interesting policy topics in coming years.

Perfect AI Deepfakes

At the moment, we cannot point to famous examples of generative AI deepfakes that fooled the public for a long period of time. They may exist and we have not surfaced the fraud yet. They are likely to exist soon.

Unless one or more governments has secret methods of differentiating AI-generated content from classically produced audio, video, and still photography, it is likely that AI content mimicking the voices of real people or generating false video from the images of living humans will create chaos and negative outcomes for the people portrayed by the AI. Even with the comparatively rudimentary AI video generating tools of the 2010s, pornographic deepfakes created problems in the lives of (primarily) women who were victimized by them. And once an audio or video file is out on the internet, victims find it difficult to erase all versions of it.

But the AI generation tools will continue to improve until it will be easy to generate a video showing Patrick Mahomes playing badminton in your backyard or Wiz Khalifa spinning discs at your office holiday party, and telling truth from fantasy will be a realm of highly-paid professionals. Will this deliver us into a post-truth world, where every digital document must be questioned and nothing can be unequivocally believed? We have always had people who bent the truth and changed the evidence to suit their own agendas. Joseph Stalin was rumored to have employed an entire building of party workers who erased people from official photographs once those photo subjects were disclaimed by Stalin's government. Soviet history was constantly being modified to meet the political demands of the next day at the whim of the leader. But when a normal person is capable of creating perfect examples of truth-bending evidence, then the whims of everyone can generate a chaotic world. The generative AI may already be good enough to birth this new world, and the competence of current generative AI is the worst it will ever be from this point forward.

As lawyers and policymakers, restricting distribution of higher-grade AI generators is one of our most powerful tools to prevent or delay this chaos. If regular people cannot have access to the ability to make perfect deepfakes, then the harms are minimized. Government can regulate the types of AI that are created and can restrict where those AI models are distributed. It can even be helpful to add costs to the deepfake machines so that they may be prohibitively expensive for consumers to acquire and use. Generative AI development companies may be convinced to place restrictions on their own products, so that the AI models will not generate realistic pictures or video of celebrities or other real people. If we believe that deepfakes are a critical societal problem, and we believe that the ability to make deepfakes for legitimate purposes is not an important society benefit, then we can justify nipping the problem at the source, and restricting the generative AI models.

Of course, those are "supply" solutions, addressing the problem by limiting effectiveness or distribution of capable AI models. We can also try "demand" solutions, as some states already have done by passing revenge porn statutes, so that, even if the competent tools exist, using them for harmful purposes is a crime. This is a more tailored method of proscribing deepfakes. People are able to use the generative AI tools as they wish, however, people who make harmful and deceptive deepfakes, or people who use deepfakes to commit fraud, are punished. People would be free to make deepfakes for their own entertainment or for parody purposes, but would not be allowed to cross the line into unlawful fraud and deception with the technology.

As these examples demonstrate, the various forms of AI in our present environment present novel problems for our society. It is important that we anticipate these problems and consider solutions with precision and practicality. Thinking like a lawyer can help us focus on real problems and stop lashing out in fear at new and complex tools. Only careful analysis can help us resolve technology problems in advance of crisis.

Final Note: Lawyers Can Use Generative AI in Compliance with Their Professional Rules

Lawyers have applied different functional types of AI for many years now. Advanced search algorithms used for legal research are predictive AI tools. Some firms have developed their own AI tools to extract helpful information from their own forms, corporate filings, and litigation materials. But no one has questioned whether these AI models can be used ethically by attorneys. This changed with the introduction of generative AI models into legal practice.

Not long after Generative Pre-trained Transformer 3 (GPT-3) burst into public consciousness in November 2022, lawyers and ethicists have called for regulation to stop lawyers from misusing the technology. But intervening time has demonstrated that the current Rules of Professional Conduct for U.S. attorneys is already broad enough to encompass and regulate a lawyer's application of generative AI to client and court work.

As described above, lawyers have violated Rule 11 of the Federal Rules of Civil Procedure and even had a law license suspended for trusting the output of generative AI to provide accurate support in court-filed documents. As the State Bar of California cautioned in ethical guidance to lawyers on generative AI, "[G]enerative AI possess the risk of encouraging greater reliance and trust on its outputs because of its purpose to generate responses and its ability to do so in a manner that projects confidence and effectively emulates human responses. A lawyer should consider these and other risks before using generative AI in providing legal services."[1] The

1. *See* The State Bar of California Standing Committee on Professional Responsibility and Conduct, Practical Guidance for the Use of Generative Artificial Intelligence in the Practice of Law.

same report also cited a lack of clarity for lawyers in how generative AI reaches and delivers its outputs.

Many lawyers do not understand that language-driven generative AI is not a search engine with drafting features, but is instead a prediction engine that simply attempts to predict—word by word—an answer to a prompt. So, the answer to your prompt is likely to contain significant errors and fabricated support. The AI model generates text that looks like legal citations without actually being legitimate legal citations. As reported in a Stanford University research study, in the first empirical evaluation of AI-driven legal research tools, the authors demonstrate that existing legal research providers' claims of accuracy are overstated. "While hallucinations [in the legal generative AI models] are reduced relative to general-purpose chatbots (GPT-4), we find that the AI research tools made by LexisNexis (Lexis+ AI) and Thomson Reuters (Westlaw AI-Assisted Research and Ask Practical Law AI) each hallucinate between 17% and 33% of the time. We also document substantial differences between systems in responsiveness and accuracy."[2] So, as of this writing, the generative AI tools used by lawyers are providing prompt responses that fail to accurately state legal principles or citations as much as one third of the time. The accuracy of these generative AI models will inevitably improve, and it is possible that these tools may have reduced hallucinations to near-zero by the time you read these words. But the professional question for lawyers is, "Can an attorney ethically use such tools, knowing about the inaccuracy?"

An answer is simple. Lawyers are responsible for the competency and honesty of their professional output. No matter how a brief, memorandum, or agreement is researched or drafted, lawyers will be held to account for the result. If we assume, for the purposes of illustration, that Microsoft Word's spellcheck program was only 80 percent reliable, missing 20 percent of a document's misspellings each time it was engaged, then would an attorney violate ethical rules in using it? No. The hypothetical semi-reliable spellcheck program could still make the drafting attorney's job quicker and the work likely better by catching even a limited number of spelling errors. However, the lawyer using this spellcheck program would still be responsible for finding the rest of the spelling errors and correcting them. The same is true for a senior litigator who asks a junior associate to

2. Magesh, Surani, Dahl, Suzgun, Manning & Ho, *Hallucination-Free? Assessing the Reliability of Leading AI Legal Research Tools*, Abstract (currently in preprint).

research and prepare a first draft of a motion for summary judgment and its support memorandum to be filed in a client's case. The draft returned as a result of this request may be perfect, but is more likely to contain flaws in logic, missed arguments, and maybe improper citations. The senior litigator, whom we assumed will be signing the filed document, is responsible for making sure the final submission to the court is correctly supported and puts forth the best arguments on behalf of the client's position. No matter what assistance a lawyer uses for document preparation, whether electronic database research, digital spelling and grammar correction, efforts of less-expensive legal minds, or generative AI models, the lawyer is still responsible under the relevant rules of professional responsibility and the applicable rules of evidence or other forum-specific requirements, to proffer an honest and competent document.

Flawed generative AI tools may be useful to lawyers despite their imperfections. But lawyers must review, research, and confirm all aspects of AI-generated content, just as the lawyer would for content drafted by an associate or paralegal. Under long-standing professional rules, lawyers are responsible for providing competent representation for clients, which means that every word an attorney releases in an official document will be used as a measure of the lawyer's competence.

The American Bar Association's (ABA) Model Rules of Professional Conduct[3] begin with a crucial foundation of legal competence (Rule 1.1), which declares, "A lawyer shall provide competent representation to a client. Competent representation requires the legal knowledge, skill, thoroughness and preparation reasonably necessary for the representation." So whatever tools may be used in legal representation, the lawyer is obligated to understand how those tools work and to use them constructively and competently on behalf of their clients. Similarly, Rule 11 of the Federal Rules of Civil Procedure[4] requires that a lawyer, by presenting signed brief or other filings to the court, certifies "that to the best of the [lawyer's]

3. I use the 2023 ABA Model Rules to illustrate requirements of the legal profession, understanding that lawyers are not regulated by the ABA Model Rules, but by the Rules of Professional Responsibility passed by the states in which those lawyers are licensed. These state rules may not perfectly match the ABA model, but they tend to include the same core concepts and to vary only on the margins.

4. Similarly reflected in the Federal Rules of Civil Procedure for other courts of law in the United States.

knowledge, information, and belief, formed after an inquiry reasonable under the circumstances:

> (1) [the filing] is not being presented for any improper purpose, such as to harass, cause unnecessary delay, or needlessly increase the cost of litigation;
> (2) the claims, defenses, and other legal contentions are warranted by existing law or by a nonfrivolous argument for extending, modifying, or reversing existing law or for establishing new law;
> (3) the factual contentions have evidentiary support or, if specifically so identified, will likely have evidentiary support after a reasonable opportunity for further investigation or discovery; and
> (4) the denials of factual contentions are warranted on the evidence or, if specifically so identified, are reasonably based on belief or a lack of information.

I present the full requirements here to emphasize that the entire litigation system operates based on the good faith and competence of lawyers filing documents to the court. Under both the professional rules and the court's procedural rules, an attorney signing a document is held to account for the assertions in the document and for the support provided those assertions. Hallucinations, however caused, would never be tolerated in a such a system, but the onus falls on the presenting counsel to weed them out, not on the court or opposing counsel to discover them (although the opposing counsel and court have incentive to find such false filings too, and they hold the filing attorney to account).

Courts constantly review filings in light of these rules and do not hesitate to hold lawyers responsible for correct filings and support. For example, in *Park v. Kim*,[5] the Second Circuit referred an attorney to the court's grievance panel for punishment after citing a fake case in a brief, stating, "The brief presents a false statement of law to this court, and it appears that attorney Lee made no inquiry, much less reasonable inquiry required by Rule 11 and long- standing precedent, into the validity of the arguments she presented." The current rules cover the content of all legal work and therefore are broad enough to encompass misuse of generative AI in the legal profession.

---

5. No. 22-2057 (2d Cir. Jan. 30, 2024).

The Fifth Circuit Court of Appeals takes a common-sense approach to the use of generative AI in filings before the court. The court's current filing cover sheet requires lawyers to certify either that no generative intelligence program was used in drafting the subject document or that "a generative artificial intelligence program was used in the drafting of this document and all generated text, including all citations and legal analysis, has been reviewed for accuracy and approved by a human." In the end, a lawyer is responsible for all content submitted to court, no matter what research or writing tools were used to produce drafts of the content.

Comment 8 to ABA Model Rule 1.1 states that "a lawyer should keep abreast of changes in the law and its practice, including the benefits and risks associated with relevant technology." Not all states have adapted a specific requirement of technical competence in their disciplinary rules, but technological understanding is simply another aspect of capable practice. A lawyer will not need the skill and knowledge to take a truck apart and put the engine back together in order to understand the risks of driving a multi-ton vehicle on the road. Similarly, if an attorney intends to use a generative AI model for work product, he will be expected to understand the primary risks and benefits involved for himself and his client. A misunderstanding of the technology can lead to problematic reliance on generative AI results, not only due to fabrications of sources and citations, but also because, as stated by the State Bar of Michigan,[6] "An algorithm may weigh factors that the law or society deem inappropriate or do so with a weight that is inappropriate in the context presented. . . . AI does not understand the world as humans do, and unless instructed otherwise, its results may reflect an ignorance of norms or case law precedent." Ignorance is a word—like intelligence—that grants too much humanization of the generative AI model. The model has no context or understanding of the underlying problems a lawyer is hired to address. It is not ignorant—ignoring the truth—the AI model simply has no understanding of what we know to be true, likely much less than a parrot which sounds like it is answering questions with human speech. The parrot may at least have an expectation of the response her sounds will elicit—make the sound for "cracker" and receive a cracker from a human. The generative AI model, no matter how sophisticated, has none.

6. JI-155, Oct. 27, 2023.

In addition to competence, lawyers are expected to provide diligent, timely work. AI may improve the speed of delivery and quality of work product when used correctly. Model Rule 1.3 states: "A lawyer shall act with reasonable diligence and promptness in representing a client." Comment 1 provides that "A lawyer must also act with commitment and dedication to the interests of the client and with zeal in advocacy upon the client's behalf." So, like the competence requirement, the zealous advocacy rule is also a catch-all obligation for lawyers that could be triggered by lazy or imprecise use of generative AI. Michael Cohen's lawyer submitted phony AI-generated citations to the court. The citations were provided by his client, but the lawyer had the responsibility under Rule 1.3 to check the citation before sending them to the court.

Generative AI is a novel tool that can be applied to many tasks in the legal field. Practicing lawyers may need to ask their clients before using new or untested AI services in representation. Model Rule 1.4 states that "A lawyer shall . . . reasonably consult with the client about the means by which the client's objectives are to be accomplished." Comment 3 of the Rule clarifies, "In some situations—depending on both the importance of the action under consideration and the feasibility of consulting with the client—this duty will require consultation prior to taking action." In Comment 5, "The client should have sufficient information to participate intelligently in decisions concerning the objectives of the representation and the means by which they are to be pursued, to the extent the client is willing and able to do so." It further states, "The guiding principle is that the lawyer should fulfill reasonable client expectations for information consistent with the duty to act in the client's best interests, and the client's overall requirements as to the character of representation." If the client should expect to know if an unusual or controversial AI tool will be used by the lawyer, that lawyer will be expected to provide relevant information about the tool. This may be especially true where courts require notification to the judge and opposing counsel of generative AI use. A client may be more interested in knowing and approving the assistance of the novel technology where the court and opposing party will soon be informed of the fact.

Both the California Ethics Opinion cited above and Florida Bar Ethics Opinion 24-1[7] state that lawyers should request and receive their clients'

7. Jan. 19, 2024.

permissions before using generative AI on client matters. California notes that lawyers should explain the risks and benefits of generative AI to clients before using the tool on a client matter. Finally, attorneys may neither use novel AI tools for their own convenience if that tool does not serve the client's interests too, nor may lawyers hide the use of these tools from their clients. According to Model Rule 1.4, Comment 7, "A lawyer may not withhold information to serve the lawyer's own interest or convenience or the interests or convenience of another person." Note that, as time passes and generative AI models become less of a novelty, it is less likely that state bar ethics counsel will be advocating for giving clients advance notice of its use in the client's case. Right now, there are controversial and uncertain aspects to applying generative AI in research and drafting. At one point, ethics counsel used to advise that lawyers specifically request client permission to employ digital research tools. This position arose less because of uncertainty about the tool, but more because a lawyer could quickly run up usage costs that the client would be expected to pay. At that stage of legal technological advancement, the time taken to look up relevant cases in law books was often less expensive and more (or equally) effective than a digital search. Now the search of databases is expected and a there is an argument that a lawyer may be committing malpractice by *not* checking digital database resources for legal research.[8] A lawyer would not expect to ask a client prior to using a word-processing program or corresponding by email, but these were once new to the legal profession and concerning for ethicists. Technology becomes more familiar and less frightening over time.

However, one of the true early problems with free public generative AI platforms is a potential loss of privacy, confidentiality, and privilege. Many generative AI models incorporate all input prompts and all output results into the training material for their models. In other words, your queries requesting a summary of your discovery material (as well as the material itself) or your request to prepare a complaint on a specific set of facts will be incorporated and possibly entirely saved within the generative AI model so that this information may be called forth in later prompts

8. Comment 8 to Rule 1.1 of the ABA Model Rules requires attorneys to remain "abreast of changes in the law and its practice, including the benefits and risks associated with relevant technology." If finding cases that support an important argument can be as simple as running a single search through a relevant publicly-available database, then a lawyer who neglects to make the argument for lack of support, and who hasn't checked the public database, may be violating Rule 1.1.

by other people. The same could be true for the generative AI output delivered to you by the model. So, if you have used these public generative AI models to assist your client's case, and the opposing party knew, or guessed, that you did so, the opposing party could conceivably use the same generative AI model to request your client's confidential information and your case strategies as entered into the model. For these reasons, legal ethicists have insisted that lawyers looking to protect client privilege and confidentiality must only use generative AI models that will not incorporate new inputs or outputs into the model to benefit future users.

The Model Rules confirm that attorneys must be careful not to reveal client confidences when using AI on the client's behalf. Rule 1.6 states, "(a) A lawyer shall not reveal information relating to the representation of a client." Comment 2, elucidates that "A fundamental principle in the client-lawyer relationship is that, in the absence of the client's informed consent, the lawyer must not reveal information relating to the representation." This includes, as stated in Comment 19 of this Rule, that "the lawyer must take reasonable precautions to prevent the information from coming into the hands of unintended recipients." Attorneys must understand that under many early generative AI platforms, there is no confidentiality for the prompts input to receive the generated output. So a prompt that includes a client name and an indication of legal problem or strategy could find its way into the wrong hands. Avoiding use of such public generative AI models is a reasonable precaution to be taken on behalf of a client because important legal privileges can be destroyed by exposure of relevant information, whether the exposure is intentional or accidental. Legal practice is rife with incidents where privileged information was carelessly exposed to the client's detriment.[9]

9. An inadvertent disclosure of privileged material will be deemed a waiver when the disclosing party has failed to take adequate precautions to maintain the confidentiality of the privileged communication. See *United States* v. *Gangi*, 1 F. Supp. 2d 256, 264 (S.D.N.Y. 1998) (holding that "[I]f a client wishes to preserve the privilege, it must treat the confidentiality of attorney-client communications like jewels if not crown jewels") (quoting *In re Sealed Case*, 877 F.2d 976, 980 (D.C. Cir. 1989)); *Fry* v. *McCall*, No. 95 Civ. 1915, 1998 WL 273035, at *3 (S.D.N.Y. May 28, 1998). The Court of Special Appeals of Maryland in *Elkton Care* v. *Quality Care Management, Inc.*, 145 Md. App. 532, 805 A.2d 1177 (Md. Ct. Spec. App. 2002) examined why some courts have adopted the "strict test" or "waiver" test under which an inadvertent disclosure constitutes a waiver of the privilege and why some others have adopted the "lenient" or "no waiver" test, under which the attorney's negligence cannot waive the privilege. The court adopted a third view and considered

Unless generative AI model providers represent otherwise, it is likely that your prompts entered into their systems are used for additional training of the AI model. It is possible that prompts and the responses they produce could be available to the general public. In this circumstance, your prompt could be revealed either by accident or by specially-designed inquiries to the same generative AI tool. A lawyer must understand where the information is going and how it may be shared before undertaking client work using AI tools. Judge Peter H. Kang of the Northern District of California in his standing discovery order requires litigants to maintain records of all "prompts or inquiries" submitted to any third-party AI tools to address confidentiality and privilege issues.[10]

Model Rule 1.18 extends the confidentiality rule to prospective clients: "Even when no client-lawyer relationship ensues, a lawyer who has learned information from a prospective client shall not use or reveal that information." So an attorney cannot be loading confidential information of client prospects into many AI models. Lawyers must understand the breadth of their confidentiality responsibilities when typing information into generative AI models, because data provided by the client may be confidential even if it relates to third parties. Certain generative AI vendors understand an attorney's obligation to keep information confidential, and therefore, the vendors selling AI tools to the legal profession offer generative AI models that do not use lawyer prompts or outputs as training material. So there is a solution to this problem, but it must be affirmatively sought by lawyers. In addition, many law firms use only "extractive AI" which generates output based only on that law firm's separate databases fed to the model.

AI could also affect billing practices. A brief written by prompting a generative AI program might take substantially less time to complete than one written directly by lawyers, even after the lawyers check the arguments

---

the following factors to be relevant: "(1) the reasonableness of the precautions taken to prevent inadvertent disclosure in view of the extent of the document production; (2) the number of inadvertent disclosures; (3) the extent of the disclosure; (4) any delay and measures taken to rectify the disclosure; and (5) whether the overriding interests of justice would or would not be served by relieving a party of its error." (*Elkton Care* v. *Quality Care Management, Inc.,*. at 545) (citing *Sampson Fire Sales* v. *Oaks*, 201 F.R.D., 351, 360 (M.D.Pa. 2001)).

10. Submitting potentially confidential and privileged information to the court raises privilege issues too.

and all citations. But will the AI-generating attorney pass this saving on to their client or charge the traditional rate for how long the brief should have taken? Pocketing the time savings may be a violation of the Model Rules if the lawyer promised to bill the clients by the hour, but does not provide the billing discounts that occurred due to AI-created efficiencies. Rule 1.5(a)(1) allows that, "A lawyer shall not make an agreement for, charge, or collect an unreasonable fee or an unreasonable amount for expenses. The factors to be considered in determining the reasonableness of a fee include . . . the time and labor required." There are also ways that AI could be used to manage billing that could violate this rule on fair charging, including allowing AI to generate bills for work that was not performed. Both the California and Florida ethical guidance do not permit lawyers to bill clients for the time saved through the use of AI. Productive use of generative AI may lead to more flat fee or retainer agreements between lawyers and clients and less pure hourly billing. An attorney must make the billing agreement before the work is performed to fairly represent the client through the use of generative AI, and then, must charge that client consistent with the billing agreement.

One of the highest impact features of generative AI is its ability to generate entirely fabricated audio and visual outputs. Some of this material generated in the style of an artist or directly incorporating previously copyrighted work will be filling our courts with copyright cases for years, as litigants push for decisions about how much similarity to older work is allowed under the fair-use doctrine. However, the most troubling aspect of this problem is called the deepfake—a fictitious audio clip, video, or still picture that proports to show a real person in a situation that the person never participated in. Generative AI can convincingly place one person's face on a made-up body or a body of another person.[11] Generative AI can

11. For example, in 2018, a deepfake video of former U.S. President Barack Obama was created by Jordan Peele and BuzzFeed. In the video, Obama appears to say things he actually never said, demonstrating the potential dangers of deepfake technology. This video was intended as a warning about the potential for misinformation and the importance of media literacy. In early 2021, a series of deepfake videos featuring actor Tom Cruise went viral on TikTok. Created by visual effects artist Chris Umé, these videos were highly realistic and showcased how convincing deepfake technology can be. The account @deeptomcruise demonstrated Cruise performing magic tricks, playing golf, and doing other activities, sparking discussions about the ethical implications of deepfake technology. In 2019, artists Bill Posters and Daniel Howe, in collaboration with advertising company Canny, created a deepfake of Facebook CEO Mark Zuckerberg. The video shows

mimic a voice and face so it seems that a person you know is talking in a language that the person doesn't speak or behaving in a manner totally foreign to that person's personality. Of course, as we have unfortunately learned over the past thirty years, new technologies will quickly be co-opted to sexually-oriented content. Some of the most infamous generative AI deepfakes have involved generating a famous person's face into a pornographic video scene. Politics is clearly another area for mischief, as generative AI has already been used as a tool to support political disinformation. The new capabilities for harmful output run across our most public and private lives.

The legal world is not immune from bad actors using generative AI to create fictions that can be offered as evidence. Lawyers and clients could misuse generative AI to create misinformation, disinformation, deepfakes, and other made-up audio, video, and illusory apparent examples of still photography. Deepfakes might help a client when offered as evidence in a divorce proceeding or a criminal case. A fake digital audio could purport to show a contractual party making statements against interest. A recent Maryland case saw a school official shunned for voicing hurtful and socially unacceptable statements, when it turned out the recordings heard by many in the community were generative AI deepfakes created by an adversary. These deepfakes mimicked the school official's voice to generate statements the official never said. In short, AI can create false evidence, that, if used at trial, contract negotiations, or in settlement discussions, could lead to serious violations of the Model Rules.[12]

The Model Rules require lawyers to be truthful in many important professional circumstances. Model Rule 3.3(a) is clear, stating, "A lawyer shall not knowingly: (1) make a false statement of fact or law to a tribunal or fail to correct a false statement of material fact or law previously made to the tribunal by the lawyer; (2) fail to disclose to the tribunal legal authority in the controlling jurisdiction known to the lawyer to be directly adverse to the position of the client and not disclosed by opposing counsel;

Zuckerberg talking about controlling the future through stolen data, highlighting concerns about data privacy, and the power of social media platforms. It was part of an art installation to raise awareness about digital surveillance. In addition, YouTube channel, *DesiFakes*, put Jerry Seinfeld onto the character of Vincent Vega in the famous "Royale with Cheese" scene.

12. Marianna Spring, *The racist AI deepfake that fooled and divided a community*, British Broadcasting Company, Oct. 4, 2024.

or (3) offer evidence that the lawyer knows to be false." Intentionally providing any false evidence violates this rule. Creating, or even knowing about, a deepfake proffered as evidence would be both offering false evidence and making a false statement of fact to the court.

Generative AI may be used to create intentionally faked evidence, like audio or video of people apparently saying words they never uttered, or it could be generating faked support statements that a lawyer accidentally posits as true. We have already seen in New York courts where a lawyer introduced a brief including AI-generated false cases, which not only violates Rule 11 of the Federal Rules of Civil Procedure, but also violates paragraph (a)(2) of Model Rule 3.3. This rule even requires correcting an earlier falsity, so the lawyer has a duty to inform the court if such an AI deepfake was introduced and the lawyer learns of its falsity later in the proceeding. We have seen that generative AI can easily provide false statements of the law in court filings, including completely fictitious cases and other fabricated supporting legal matter, and using this material on behalf of a client would violate the legal disciplinary rules as false presentations of the law.

Similarly, under Model Rule 3.3(a)(3), a lawyer must refuse to allow offers of false evidence by testimony. AI-generated deepfakes could serve to underly the false testimony of a witness, and an attorney violates the disciplinary rules by offering such testimony. Rule 3.3(a)(3) states, "If a lawyer, the lawyer's client, or a witness called by the lawyer, has offered material evidence and the lawyer comes to know of its falsity, the lawyer shall take reasonable remedial measures, including, if necessary, disclosure to the tribunal. A lawyer may refuse to offer evidence, other than the testimony of a defendant in a criminal matter, that the lawyer reasonably believes is false." Under Comment 3, "An advocate is responsible for pleadings and other documents prepared for litigation." Comment 5 explains that the rule "requires that the lawyer refuse to offer evidence that the lawyer knows to be false, regardless of the client's wishes. This duty is premised on the lawyer's obligation as an officer of the court to prevent the trier of fact from being misled by false evidence." So, from this point forward, where a client produces audio, video, or pictorial evidence to support his position, a lawyer should be asking how the evidence was produced and if it can be authenticated. Willful blindness to generative AI deepfaked evidence will prove harmful to lawyers who have reasons to question that evidence before presenting it to a court.

Further, even outside of the court context, lawyers are obligated to avoid false statements, which would include AI-generated deepfakes. Rule 4.1(a) provides, "In the course of representing a client a lawyer shall not knowingly . . . make a false statement of material fact or law to a third person." Comment 1 states, "A misrepresentation can occur if the lawyer incorporates or affirms a statement of another person that the lawyer knows is false. Misrepresentations can also occur by partially true but misleading statements or omissions that are the equivalent of affirmative false statements." And the obligation goes further as illustrated in Comment 3 states, "a lawyer is prohibited from counseling or assisting a client in conduct that the lawyer knows is criminal or fraudulent." So a lawyer may not help a client create or otherwise use AI-generated deepfakes even outside the bounds of court. This is also supported by Rule 8.4(c) which identifies professional misconduct as engaging "in conduct involving dishonesty, fraud, deceit or misrepresentation." So where a client or a client's agent has used generative AI to create deepfakes or other misinformation, that client's lawyer has an obligation to question the authenticity of the evidence before such evidence is offered to a court or even in settlement negotiations with an opposing party. As generative AI-produced deepfakes become easier to create and harder to disprove, lawyers will need to take additional care in using evidence provided by clients and others. Case law in the next years may show courts and state bar ethics councils strengthening obligations for attorneys to investigate written, audio, photographic, and video evidence provided by clients. Evidentiary fraud may create the next realm of generative AI-created public professionalism cases for lawyers.

The Model Rules prohibit lawyers from allowing non-ethical behavior of their agents and junior attorneys, and confer an affirmative duty on lawyers to supervise the professional conduct of employees and agents. This would include the above-described misuse of AI. Rule 5.1(b) provides that "A lawyer having direct supervisory authority over another lawyer shall make reasonable efforts to ensure that the other lawyer conforms to the Rules of Professional Conduct." Also Rule 5.1(c) states, "A lawyer shall be responsible for another lawyer's violation of the Rules of Professional Conduct if:

(1) the lawyer orders or, with knowledge of the specific conduct, ratifies the conduct involved; or

(2) the lawyer is a partner or has comparable managerial authority in the law firm in which the other lawyer practices, or has direct supervisory authority over the other lawyer, and knows of the conduct at a time when its consequences can be avoided or mitigated but fails to take reasonable remedial action."

This should be relatively clear—the senior attorney has the responsibility for work product and behavior of her team. See Comment 2 of Rule 5.1, which clarifies that this Rule "requires lawyers with managerial authority within a firm to make reasonable efforts to establish internal policies and procedures designed to provide reasonable assurance that all lawyers in the firm will conform to the Rules of Professional Conduct." Comment 8 adds, "The duties imposed by this Rule on managing and supervising lawyers do not alter the personal duty of each lawyer in a firm to abide by the Rules of Professional Conduct." So if the younger lawyers are experimenting with generative AI for work products, the senior lawyers at the firm are responsible for understanding if this is negatively affecting work product or violating the Model Rules cited above. Both the junior and senior counsel are responsible for the end product. Rule 5.3 extends this responsibility to non-lawyer agents as well. In regard to generative AI deepfakes, lawyers are responsible for their use, including use by clients and/or agents. A new era of evidence review is upon us, and as deceptive generative AI output becomes easier to produce and harder to disprove, the lawyer's obligations to review the evidence of both sides become more important.

The ABA Model Rules for Professional Conduct provide all the regulation and guidance needed for lawyers to ethically utilize generative AI to represent client interests. The ABA Model Rules of Professional Responsibility may not have been written to address generative AI, but the Rules are comprehensive enough to enforce responsible generative AI use by lawyers. Unless or until different and unforeseen generative AI issues arise, no additional rules need to be promulgated for the ethical use of these tools in legal practice.[13] The onus falls on lawyers to confirm the

13. Judge Paul W. Grimm of Duke University Law School and Maura R. Grossman of University of Waterloo Law School have proposed an amendment to Rule 901(b) of Federal Rules of Evidence (1975), so that, to satisfy the requirement of authenticating or identifying an item of evidence, the proponent must produce evidence sufficient to support a finding that the item is what the proponent claims it is. This evidentiary rule, a change in which may or may not benefit the courts, is distinct from the professional ethics rules that govern law practice in the states.

accuracy and effectiveness of their own production, whether or not the latest tools have assisted the development of legal product.

This also means that lawyers are free to access generative AI models if the models are useful to client representation and can be protective of client confidences. Generative AI models, like most new technology, can be applied where attorneys understand the risks and can manage those risks.

Index

About the Author

Ted Claypoole is a digital technology lawyer, writer, and editor. He is a partner and team leader for Womble Bond Dickinson, a regular speaker and thought leader on AI policy, and father of three brilliant daughters. His other books include, *The Law of Artificial Intelligence and Smart Machines*, *Privacy in the Age of Big Data*, and the crime novel *False Consolation.*